Preface

Students commonly find difficulty with problems in fluid mechanics. They may misunderstand what is required or misapply the solutions. This book is intended to help. It is a collection of problems in elementary fluid mechanics with accompanying solutions, and intended principally as a study aid for undergraduate students of chemical engineering — although students of all engineering disciplines will find it useful. It helps in preparation for examinations, when tackling coursework and assignments, and later in more advanced studies of the subject. In preparing this book I have not tried to replace other, fuller texts on the subject. Instead I have aimed at supporting undergraduate courses and academic tutors involved in the supervision of design projects.

In the text, worked examples enable the reader to become familiar with, and to grasp firmly, important concepts and principles in fluid mechanics such as mass, energy and momentum. The mathematical approach is simple for anyone with prior knowledge of basic engineering concepts. I have limited the problems to those involving incompressible, Newtonian fluids and single-phase flow through pipes. There is no attempt to include the effects of compressible and non-Newtonian fluids, or of heat and mass transfer. I also held back from more advanced mathematical tools such as vectorial and tensorial mathematics.

Many of the problems featured have been provided by university lecturers who are directly involved in teaching fluid mechanics, and by professional engineers in industry. I have selected each problem specifically for the light it throws on the fundamentals applied to chemical engineering, and for the confidence its solution engenders.

The curricula of university chemical engineering degree courses cover the fundamentals of fluid mechanics with reasonable consistency although, in certain areas, there are some differences in both procedures and nomenclature. This book adopts a consistent approach throughout which should be recognizable to all students and lecturers.

I have tailored the problems kindly contributed by industrialists to safeguard commercial secrets and to ensure that the nature of each problem is clear. There is no information or detail which might allow a particular process or company to be recognized. All the problems use SI units. As traditional systems of units are still very much in use in industry, there is a table of useful conversions. Fluid mechanics has a jargon of its own, so I have included a list of definitions.

There are nine chapters. They cover a range from stationary fluids through fluids in motion. Each chapter contains a selected number of problems with solutions that lead the reader step by step. Where appropriate, there are problems with additional points to facilitate a fuller understanding. Historical references to prominent pioneers in fluid mechanics are also included. At the end of each chapter a number of additional problems appear; the aim is to extend the reader's experience in problem-solving and to help develop a deeper understanding of the subject.

I would like to express my sincere appreciation to Dr Robert Edge (formerly of Strathclyde University), Mr Brendon Harty (Roche Products Limited), Dr Vahid Nassehi (Loughborough University), Professor Christopher Rielly (Loughborough University), Professor Laurence Weatherley (University of Canterbury), Dr Graeme White (Heriot Watt University), Mr Martin Tims (Esso UK plc) and Miss Audra Morgan (IChemE) for their assistance in preparing this book. I am also grateful for the many discussions with professional engineers from ICI, Esso and Kvaerner Process Technology.

The text has been carefully checked. In the event, however, that readers uncover any error, misprint or obscurity, I would be grateful to hear about it. Suggestions for improvement are also welcome.

Carl Schaschke
April 2000

List of symbols

The symbols used in the worked examples are defined below. Where possible, they conform to consistent usage elsewhere and to international standards. SI units are used although derived SI units or specialist terms are used where appropriate. Specific subscripts are defined separately.

Roman	Term	SI or preferred unit
a	area of pipe or orifice	m^2
A	area of channel or tank	m^2
B	breadth of rectangular weir	m
c	constant	—
c	velocity of sound	ms^{-1}
C	Chézy coefficient	$m^{1/2}s^{-1}$
C	coefficient	—
C	concentration	gl^{-1}
d	diameter	m
D	impeller diameter	m
f	fraction	—
f	friction factor	—
F	depth of body below free surface	m
F	force	N
g	gravitational acceleration	ms^{-2}
H	head	m
i	slope of channel	—
k	constant	—
L	fundamental dimension for length	—
L	length	m
L	mass loading	$kgm^{-2}s^{-1}$
m	mass	kg
m	mass flowrate	kgs^{-1}
m	mean hydraulic depth	m
M	fundamental dimension for mass	—
n	channel roughness	$m^{-1/3}s$

n	number of pipe diameters	—
N	rotational speed	rps
N_s	specific speed	$m^{3/4}s^{-3/2}$
p	pressure	Nm^{-2}
P	power	W
P	wetted perimeter	m
Q	volumetric flowrate	m^3s^{-1}
r	radius	m
R	frictional resistance	Nm^{-2}
R	radius	m
s	depth	m
S_n	suction specific speed	—
t	thickness of oil film	mm
t	time	s
T	fundamental dimension for time	—
T	torque	Nm
v	velocity	ms^{-1}
V	volume	m^3
W	width	m
W	work	W
x	principal co-ordinate	—
x	distance	m
y	principal co-ordinate	—
y	distance	m
z	principal co-ordinate	—
z	static head	m

Greek

β	ratio of pipe to throat diameter	—
δ	film thickness	mm
Δ	finite difference	—
ε	absolute roughness	mm
η	pump efficiency	—
θ	angle	
λ	friction factor	—
μ	dynamic viscosity	Nsm^{-2}
ν	kinematic viscosity	m^2s^{-1}
π	3.14159	
ρ	density	kgm^{-3}
σ	surface tension	Nm^{-1}
τ	shear stress	Nm^{-2}
ϕ	friction factor	—
ω	angular velocity	radians s^{-1}

Fluid mechanics and problem-solving

Fluid mechanics forms an integral part of the education of a chemical engineer. The science deals with the behaviour of fluids when subjected to changes of pressure, frictional resistance, flow through pipes, ducts, restrictions and production of power. It also includes the development and testing of theories devised to explain various phenomena. To the chemical engineer, a knowledge of the behaviour of fluids is of crucial importance in cost-effective design and efficient operation of process plant.

Fluid mechanics is well known for the large number of concepts needed to solve even the apparently simplest of problems. It is important for the engineer to have a full and lucid grasp of these concepts in order to attempt to solve problems in fluid mechanics. There is, of course, a considerable difference between studying the principles of the subject for examination purposes, and their application by the practising chemical engineer. Both the student and the professional chemical engineer, however, require a sound grounding. It is essential that the basics are thoroughly understood and can be correctly applied.

Many students have difficulty in identifying relevant information and fundamentals, particularly close to examination time. Equally, students may be hesitant in applying theories covered in their studies, resulting from either an incomplete understanding of the principles or a lack of confidence caused by unfamiliarity. For those new to the subject, finding a clear path to solving a problem may not always be straightforward. For the unwary and inexperienced, the opportunity to deviate, to apply incorrect or inappropriate formulae or to reach a mathematical impasse in the face of complex equations, is all too real. The danger is that the student will dwell on a mathematical quirk which may be specific purely to the manner in which the problem has been (incorrectly) approached. A disproportionate amount of effort will therefore be expended on something irrelevant to the subject of fluid mechanics.

Students develop and use methods for study which are dependent on their own personal needs, circumstances and available resources. In general, however, a quicker and deeper understanding of principles is achieved when a problem is provided with an accompanying solution. The worked example is a recognized and widely-used approach to self-study, providing a clear and logical approach from a distinct starting point through defined steps, together with the relevant mathematical formulae and manipulation. This method benefits the student by appreciation of both the depth and complexity involved in reaching a solution.

While some problems in fluid mechanics are straightforward, unexpected difficulties can be encountered when seemingly similar or related simple problems require the evaluation of a different but associated variable. Although the solution may require the same starting point, the route through to the final answer may be quite different. For example, determining the rate of uniform flow along an inclined channel given the dimensions of the channel is straightforward. But determining the depth of flow along the channel for given parameters in the flow presents a problem. Whereas the former is readily solved analytically, the latter is complicated by the fact that the fluid velocity, flow area and a flow coefficient all involve the depth of flow. An analytical solution is no longer possible, thus requiring the use of graphical or trial and error approaches.

There are many similarities between the governing equations in heat, mass and momentum transport and it is often beneficial to bring together different branches of the subject. Other analogies between different disciplines are also useful, although they must be applied with care. In fluid mechanics, analogies between electrical current and resistance are often used, particularly in dealing with pipe networks where the splitting and combining of lines can be likened to resistors in parallel and in series.

Some applications of fluid mechanics require involved procedures. Selecting a pump, for example, follows a fairly straightforward set of well-defined steps although the lengthy procedure needed can become confusing. It is important to establish the relationship between the flowrate and pressure, or head, losses in the pipework connecting process vessels together. With frictional losses due to pipe bends, elbows and other fittings represented by either equivalent length pipe or velocity heads, pumping problems therefore require careful delineation. Any pump calculation is best reduced to the evaluation of the suction pressure or head and then of the discharge head; the difference is the delivery head required from the pump. For a sizing calculation, all that is really needed is to determine the delivery head for the required volumetric flowrate. As in many process engineering calculations dealing with equipment sizing, the physical layout plays an important part, not only in standardizing the method for easy checking but also in simplifing the calculations. Obviously there will be cases requiring more detail but, with a bit of attention, such deviations from practice can easily be incorporated.

Finally, the application of fluid mechanics in chemical engineering today relies on the fundamental principles largely founded in the seventeenth and eighteenth centuries by scientists including Bernoulli, Newton and Euler. Many of today's engineering problems are complex, non-linear, three-dimensional and transient, requiring interdisciplinary approaches to solution. High-speed and powerful computers are increasingly used to solve complex problems, particularly in computational fluid dynamics (CFD). It is worth remembering, however, that the solutions are only as valid as the mathematical models and experimental data used to describe fluid flow phenomena. There is, for example, no analytical model that describes precisely the random behaviour of fluids in turbulent motion. There is still no substitute for an all-round understanding and appreciation of the underlying concepts and the ability to solve or check problems from first principles.

Contents

‘The scientist describes what is:
the engineer creates what never was.’
Theodore von Kármán (1881–1963)

‘I hear, and I forget
I see, and I remember
I do, and I understand.’
Anonymous

1 Fluid statics

Introduction

Fluids, whether moving or stationary, exert forces over a given area or surface. Fluids which are stationary, and therefore have no velocity gradient, exert normal or pressure forces whereas moving fluids exert shearing forces on the surfaces with which they are in contact. It was the Greek thinker Archimedes (c287BC–c212BC) who first published a treatise on floating bodies and provided a significant understanding of fluid statics and buoyancy. It was not for another 18 centuries that the Flemish engineer Simon Stevin (1548–1620) correctly provided an explanation of the basic principles of fluid statics. Blaise Pascal (1623–1662), the French mathematician, physicist and theologian, performed many experiments on fluids and was able to illustrate the fundamental relationships involved.

In the internationally accepted SI system (Système International d'Unités), the preferred derived units of pressure are Newtons per square metre (Nm^{-2}) with base units of $kgm^{-1}s^{-1}$. These units, also known as the Pascal (Pa), are relatively small. The term bar is therefore frequently used to represent one hundred thousand Newtons per square metre (10^5 Nm^{-2} or 0.1 MPa). Many pressure gauges encountered in the process industries are still to be found calibrated in traditional systems of units including the Metric System, the Absolute English System and the Engineers' English System. This can lead to confusion in conversion although many gauges are manufactured with several scales. Further complication arises since the Pascal is a relatively small term and SI recommends that any numerical prefix should appear in the numerator of an expression. Although numerically the same, Nmm^{-2} is often wrongly used instead of MNm^{-2}.

It is important to note that the pressure of a fluid is expressed in one of two ways. Absolute pressure refers to the pressure above total vacuum whereas gauge pressure refers to the pressure above atmospheric, which itself is a variable quantity and depends on the local meteorological conditions. The atmospheric pressure used as standard corresponds to 101.3 kNm^{-2} and is equivalent to approximately 14.7 pounds force per square inch, or a barometric reading of 760 mmHg. The pressure in a vacuum, known as absolute zero, therefore corresponds to a gauge pressure of -101.3 kNm^{-2} assuming standard atmospheric pressure. A negative gauge pressure thus refers to a pressure below atmospheric.

The barometer is a simple instrument for accurately measuring the atmospheric pressure. In its simplest form it consists of a sealed glass tube filled with a liquid (usually mercury) and inverted in a reservoir of the same liquid. The atmospheric pressure is therefore exerted downwards on the reservoir of liquid such that the liquid in the tube reaches an equilibrium elevation. Above the liquid meniscus exists a vacuum, although in actual fact it corresponds to the vapour pressure of the liquid. In the case of mercury this is a pressure of 10 kNm^{-2} at 20°C.

In addition to gauges that measure absolute pressure, there are many devices and instruments that measure the difference in pressure between two parts in a system. Differential pressure is of particular use for determining indirectly the rate of flow of a process fluid in a pipe or duct, or to assess the status of a particular piece of process equipment during operation — for example, identifying the accumulation of deposits restricting flow, which is important in the case of heat exchangers and process ventilation filters.

Although there are many sophisticated pressure-measuring devices available, manometers are still commonly used for measuring the pressure in vessels or in pipelines. Various forms of manometer have been designed and generally are either open (piezometer) or closed (differential). For manometer tubes with a bore of less than 12 mm, capillary action is significant and may appreciably raise or depress the meniscus, depending on the manometric fluid.

Finally, while fluids may be described as substances which offer no resistance to shear and include both gases and liquids, gases differ from liquids in that they are compressible and may be described by simple gas laws. Liquids are effectively incompressible and for most practical purposes their density remains constant and does not vary with depth (hydrostatic pressure). At ultra high pressures this is not strictly true. Water, for example, has a 3.3% compressibility at pressures of 69 MNm^{-2} which is equivalent to a depth of 7 km. It was Archimedes who first performed experiments on the density of solids by immersing objects in fluids. The famous story is told of Archimedes being asked by King Hiero to determine whether a crown was pure gold throughout or contained a cheap alloy, without damaging the crown. Supposedly, while in a public bath, Archimedes is said to have had a sudden thought of immersing the crown in water and checking its density. He was so excited that he ran home through the streets naked shouting 'Eureka! Eureka! — I have found it! I have found it!'.

1.1 Pressure at a point

Determine the total force on a wall of an open tank 2 m wide containing fuel oil of density 924 kgm^{-3} at a depth of 2 m.

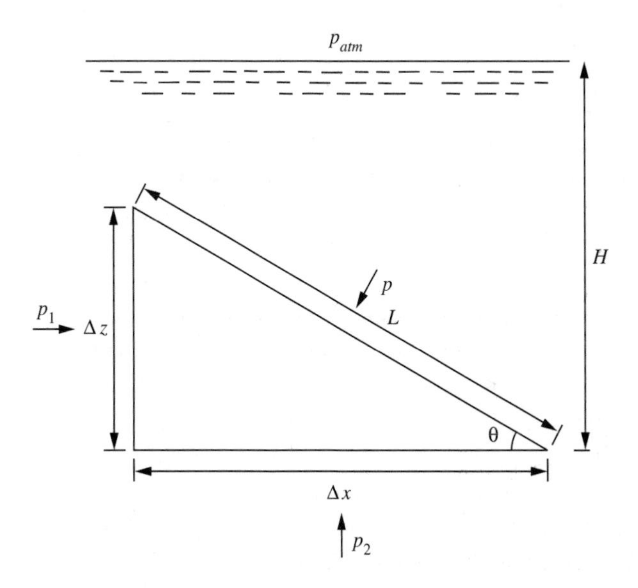

Solution

To determine the pressure at a point in the static liquid below the free surface, consider the equilibrium forces on a wedge-shaped element of the liquid. Resolving in the x-direction

$$p \Delta y L \sin\theta - p_1 \Delta y \Delta z = 0$$

where

$$\sin\theta = \frac{\Delta z}{L}$$

Then

$$p = p_1$$

Resolving in the z-direction

$$F + p \Delta y L \cos\theta - p_2 \Delta x \Delta y = 0$$

where

$$\cos\theta = \frac{\Delta x}{L}$$

and the weight (force due to gravity) of the element is

$$F = mg$$

$$= \rho \frac{\Delta x \Delta z}{2} \Delta y g$$

If the element is reduced to zero size, in the limit this term disappears because it represents an infinitesimal higher order than the other terms and may be ignored. Thus

$$p = p_2$$

Note that the angle of the wedge-shaped element is arbitrary. The pressure p is therefore independent of θ. Thus, the pressure at a point in the liquid is the same in all directions (Pascal's law). To determine the pressure at a depth H, the equilibrium (upward and downward) forces are

$$p_{atm} \Delta x \Delta y + \rho \Delta x \Delta y H g - p \Delta x \Delta y = 0$$

which reduces to

$$p = p_{atm} + \rho g H$$

The pressure (above atmospheric) at the base of the tank is therefore

$$p = \rho g H$$

$$= 924 \times g \times 2$$

$$= 18.129 \times 10^3 \text{ Nm}^{-2}$$

The total force exerted over the wall is therefore

$$F = \frac{pa}{2}$$

$$= \frac{18.129 \times 10^3 \times 2 \times 2}{2}$$

$$= 36.258 \times 10^3 \text{ N}$$

The total force is found to be 36.26 kNm^{-2}.

1.2 Pressure within a closed vessel

A cylindrical vessel with hemispherical ends is vertically mounted on its axis. The vessel contains water of density 1000 kgm^{-3} and the head space is pressurized to a gauge pressure of 50 kNm^{-2}. The vertical wall section of the vessel has a height of 3 m and the hemispherical ends have radii of 1 m. If the vessel is filled to half capacity, determine the total force tending to lift the top dome and the absolute pressure at the bottom of the vessel.

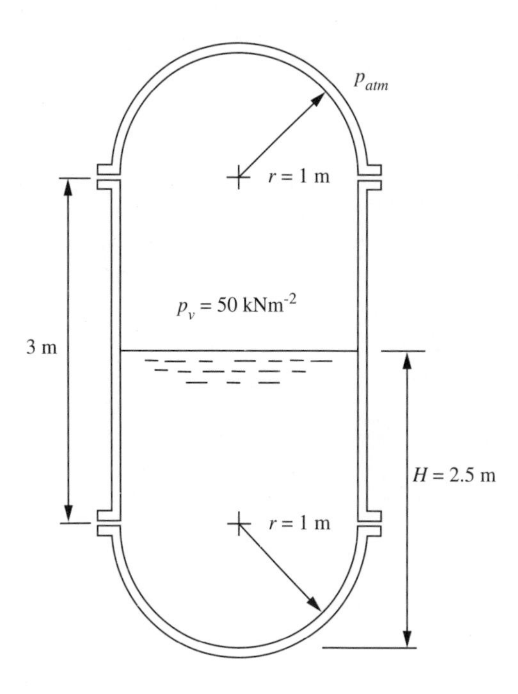

Solution

The total vertical force, F, tending to lift the dome is the pressure applied over the horizontal projected area

$$F = p_v \pi r^2$$

where p_v is the gauge pressure within the vessel. That is

$$F = 50 \times 10^3 \times \pi \times 1^2$$

$$= 1.56 \times 10^6 \text{ N}$$

Note that above the liquid surface the pressure in the head space is exerted uniformly on the inner surface of the vessel. Below the liquid, however, the pressure on the vessel surface varies with depth. The absolute pressure (pressure above a vacuum) at the bottom of the vessel is therefore

$$p = p_{atm} + p_v + \rho g H$$

$$= 101.3 \times 10^3 + 50 \times 10^3 + 1000 \times g \times 2.5$$

$$= 175.3 \times 10^3 \text{ Nm}^{-2}$$

The force tending to lift the dome is 1.56 MN and the pressure at the bottom of the vessel is 175.3 kNm^{-2}.

Note that, unlike the gas pressure which is exerted uniformly in the head space, the analysis to determine the hydrostatic forces acting on the submerged curved surface (lower domed section) requires resolving forces in both the vertical and horizontal directions. The magnitude of the horizontal reaction on the curved surface is equal to the hydrostatic force which acts on a vertical projection of the curved surface, while the magnitude of the vertical reaction is equal to the sum of the vertical forces above the curved surface and includes the weight of the liquid. In this case, however, the vessel is symmetrical such that the hydrostatic force is in the downward direction. The downward force imposed by the gas and liquid is thus

$$F = (p_v + \rho g h)\pi r^2 + \rho g \frac{2\pi r^2}{3}$$

$$= (50{,}000 + 1000 \times g \times 1.5) \times \pi \times 1^2 + 1000 \times g \times \frac{2 \times \pi \times 1^3}{3}$$

$$= 223{,}854 \text{ N}$$

$$= 223.8 \text{ kN}$$

1.3 Forces within a hydraulic ram

A hydraulic ram consists of a weightless plunger of cross-sectional area 0.003 m^2 and a piston of mass 1000 kg and cross-sectional area 0.3 m^2. The system is filled with oil of density 750 kgm^{-3}. Determine the force on the plunger required for equilibrium if the plunger is at an elevation of 2 m above the piston.

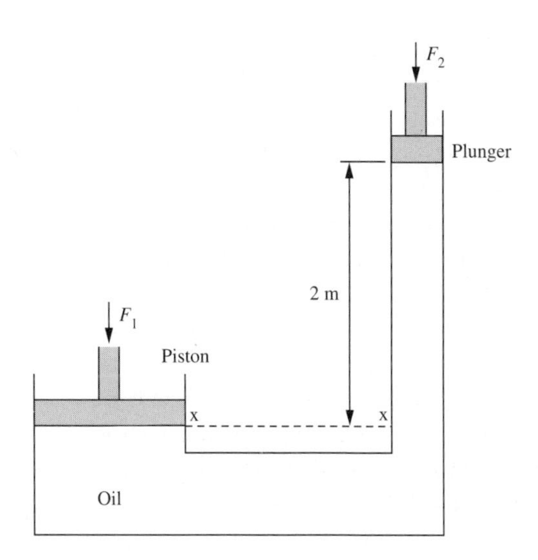

Solution

For the piston, the pressure at the datum elevation xx is

$$p_{xx} = \frac{F_1}{a_1}$$

where F_1 is the force of the piston and a_1 is the area of the piston. This pressure is equal to the pressure applied by the plunger at the same datum elevation. That is

$$p_{xx} = \frac{F_2}{a_2} + \rho_{oil} gH$$

where F_2 is the force on the plunger, a_2 is the area of the plunger and H is the elevation of the plunger above the datum. Therefore

$$\frac{F_1}{a_1} = \frac{F_2}{a_2} + \rho_{oil} gH$$

Rearranging

$$F_2 = a_2\left(\frac{F_1}{a_1} - \rho_{oil} gH\right)$$

$$= 0.003 \times \left(\frac{1000 \times g}{0.3} - 750 \times g \times 2\right)$$

$$= 54\,\mathrm{N}$$

The force required for equilibrium is found to be 54 N. Note that if no downward force is applied to the weightless plunger, the plunger would rise to an elevation of 4.44 m.

The hydraulic ram illustrated is an example of a closed system in which the pressure applied by the piston is transmitted throughout the hydraulic fluid (oil). The principle of pressure transmission is known as Pascal's law after Pascal who first stated it in 1653. Hydraulic systems such as rams, lifts and jacks are based on this principle and are useful for lifting and moving purposes. It is usual in such hydraulic systems to replace the piston with compressed air. The force applied is then controlled by the applied air pressure. High pressures can therefore be achieved, as in the case of hydraulic presses, in which the force exerted against a piston in turn exerts the force over a smaller area. For example, the plunger shown corresponds to a diameter of 62 mm over which an equilibrium pressure of 18 $\mathrm{kNm^{-2}}$ is applied. If it were to be connected to a shaft 18 mm in diameter, then the force exerted over the area of the shaft would correspond to 222 $\mathrm{kNm^{-2}}$ — a factor of 12 times greater.

1.4 Liquid-liquid interface position in a solvent separator

Mixtures which contain two mutually insoluble organic and aqueous liquids are to be separated in a separator which consists of a vertical chamber with overflow and underflow. The mixture is fed slowly to the separator in which the aqueous phase, of constant density 1100 kgm^{-3}, is discharged from the underflow at the base of the chamber to a discharge point 50 cm below the overflow level in the chamber. The organic phase can vary in density from 600–800 kgm^{-3}. Determine the minimum height of the chamber, H, which can be used if the organic phase is not to leave with the aqueous phase. If the height H is made equal to 3 m, determine the lowest possible position of the interface in the chamber below the overflow.

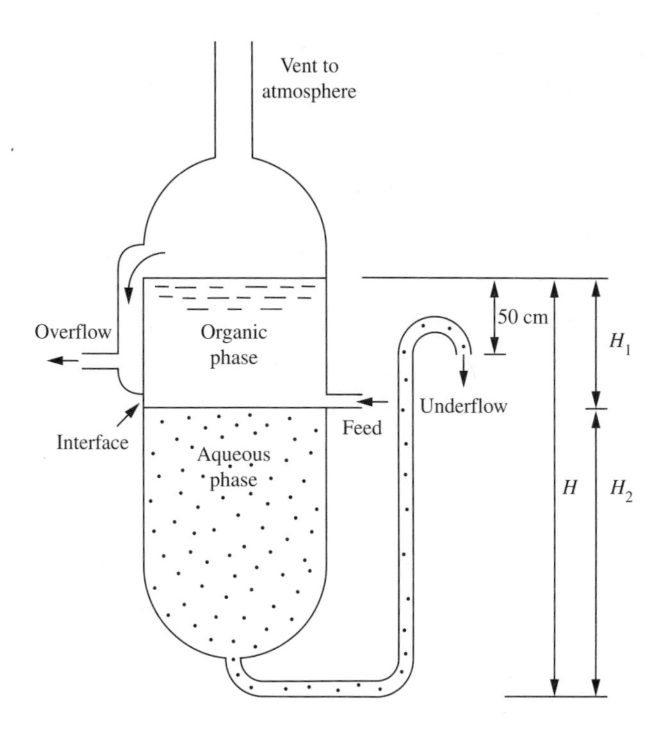

Solution

The separator is assumed to operate at atmospheric pressure. Equating the pressure in the chamber and discharge point for the maximum possible depth (in metres) for the organic phase in the chamber gives

$$\rho_o g H = \rho_{aq} g(H - 0.5)$$

where ρ_o and ρ_{aq} are the densities of organic and aqueous solutions, respectively. Rearranging

$$H = \frac{0.5\rho_{aq}}{\rho_{aq} - \rho_o}$$

To ensure no loss of organic phase with the aqueous phase, the height of the chamber is greatest for the highest possible organic density (800 kgm^{-3}). Therefore

$$H = \frac{0.5 \times 1100}{1100 - 800}$$

$$= 2.2\,\text{m}$$

For a fixed length of chamber of 3 m, the interface between the two phases is determined from the pressure in the chamber and discharge point. That is

$$\rho_o g H_1 + \rho_{aq} g H_2 = \rho_{aq} g(H - 0.5)$$

where

$$H = H_1 + H_2$$

Therefore

$$\rho_o g H_1 + \rho_{aq} g(H - H_1) = \rho_{aq} g(H - 0.5)$$

Rearranging, the interface position is at its lowest position when the organic phase has a density of 600 kgm^{-3}. That is

$$H_1 = \frac{0.5\rho_{aq}}{\rho_{aq} - \rho_o}$$

$$= \frac{0.5 \times 1100}{1100 - 600}$$

$$= 1.1\,\text{m}$$

The maximum depth is found to be 2.2 m and the interface below the overflow is found to be 1.1 m. Ideally, the feed point to the chamber should be located at the liquid-liquid interface to ensure quick and undisturbed separation. Where the density of the organic phase is expected to vary, either an average position or the position corresponding to the most frequently encountered density may be used.

1.5 Liquid-liquid interface measurement by differential pressure

Aqueous nitric acid is separated from an insoluble oil in a vessel. Dip legs extend into both phases through which air is gently discharged sufficient to overcome the hydrostatic pressure. Determine the position of the interface between the legs if the legs are separated a distance of 1 m for which the differential pressure between the legs is 10 kNm^{-2}. The densities of oil and nitric acid are 900 kgm^{-3} and 1070 kgm^{-3}, respectively.

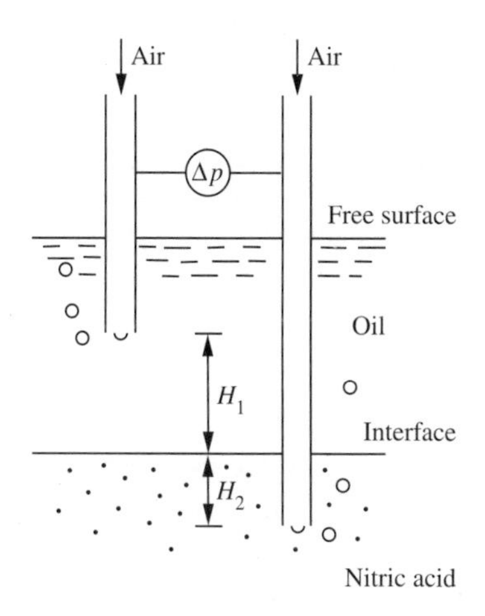

Solution

The use of dip legs is an effective way of measuring liquid densities, liquid-liquid interface positions and detecting the presence of solid material in liquids. As it has no moving or mechanical parts it is essentially maintenance free and it has therefore found application in the nuclear industry amongst others. In this application, the dip legs are used to determine the position of the liquid-liquid interface in which the densities of the two phases are assumed to be constant. The differential pressure between the legs is

$$\Delta p = \rho_o g H_1 + \rho_n g H_2$$

where ρ_o and ρ_n are the densities of the oil and nitric acid and where the fixed distance between the ends of the dip legs is

$$\begin{aligned} H &= H_1 + H_2 \\ &= 1 \text{ m} \end{aligned}$$

Eliminating H_2 and rearranging

$$H_1 = \frac{\Delta p - \rho_n g H}{(\rho_o - \rho_n) g}$$

$$= \frac{10 \times 10^3 - 1070 \times g \times 1}{(900 - 1070) \times g}$$

$$= 0.30\,\text{m}$$

The depth is found to be 30 cm below the upper dip leg.

Note that a single dip leg can be used to determine the depth of liquids of constant density in vessels in which the gas pressure applied is used to overcome the hydrostatic pressure. For cases in which the density of the liquid is likely to vary, due to changes in concentration or the presence of suspended solids, the density can be determined using two dip legs of different length, the ends of which are a fixed distance apart. In the more complicated case of two immiscible liquids in which the densities of both phases may vary appreciably, it is possible to determine the density of both phases and the location of the interface using four dip legs with two in each phase.

In practice, it is necessary to adjust carefully the gas pressure until the hydrostatic pressure is just overcome and gas flows freely from the end of the dip leg(s). Sensitive pressure sensing devices are therefore required for the low gauge pressures involved. Fluctuating pressure readings are usually experienced, however, as the gas bubbles form and break off the end of the leg. Conversion charts may then be used to convert a mean pressure reading to concentration, interface position or liquid volume, as appropriate.

1.6 Measurement of crystal concentration by differential pressure

The concentration of sodium sulphate crystals in a liquid feed to a heat exchanger is determined by a differential pressure measurement of the saturated liquid in the vertical leg feeding the heat exchanger. If the pressure measurements are separated by a vertical distance of 1.5 m, determine the density of the solution with crystals and the fraction of crystals for a differential pressure measurement of 22 kNm^{-2}*. The density of saturated sodium sulphate is 1270 kgm*$^{-3}$ *and density of anhydrous solution sulphate is 2698 kgm*$^{-3}$*.*

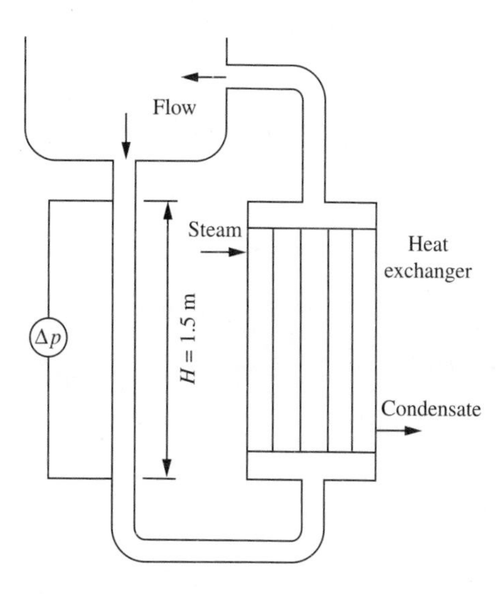

Solution

Assuming no differential pressure loss due to friction in the leg, the differential pressure is due to the static pressure between the pressure measurement points. That is

$$\Delta p = \rho g H$$

where ρ is the density of the solution.

Rearranging

$$\rho = \frac{\Delta p}{gH}$$

$$= \frac{22 \times 10^3}{g \times 1.5}$$

$$= 1495 \text{ kgm}^{-3}$$

The density of the solution with crystals is 1495 kgm^{-3}. This density is greater than that of the saturated sodium sulphate solution alone and therefore indicates the presence of crystals for which the fractional content is found from

$$\rho = f_1\rho_s + f_2\rho_c$$

where ρ_s is the density of saturated solution, ρ_c is the density of crystals, and f_1 and f_2 are the respective fractions where

$$f_1 + f_2 = 1$$

Eliminating f_1

$$f_2 = \frac{\rho - \rho_s}{\rho_c - \rho_s}$$

$$= \frac{1495 - 1270}{2698 - 1270}$$

$$= 0.157$$

That is, the crystal content is found to be 15.7%. This is, however, an overestimate since frictional effects of the flowing liquid in the leg are ignored. Where they can not be ignored the differential pressure is modified to

$$\Delta p = \rho g(H - H_L)$$

where H_L is the head loss due to friction.

1.7 Pressure within a gas bubble

A small gas bubble rising in an open batch fermenter has a radius of 0.05 cm when it is 3 m below the surface. Determine the radius of the bubble when it is 1 m below the surface. It may be assumed that the pressure inside the bubble is 2 σ/r above the pressure outside the bubble, where r is the radius of the bubble and σ is the surface tension of the gas-fermentation broth and has a value of 0.073 Nm^{-1}. The pressure and volume of the gas in the bubble are related by the expression pV = c where c is a constant.

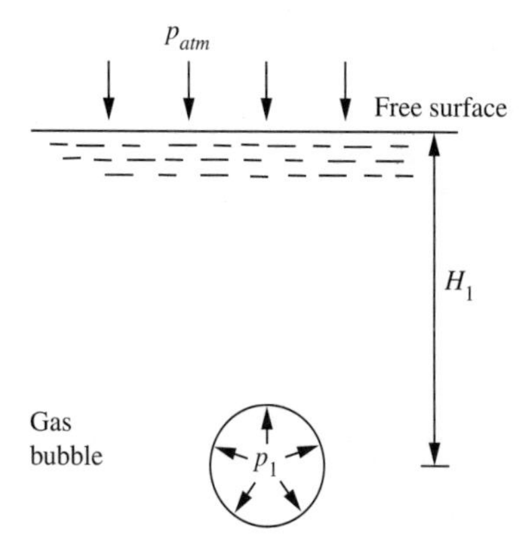

Solution

At a depth of 3 m, the pressure within the bubble, p_1, is dependent on the pressure at the free surface, the hydrostatic pressure and surface tension effect. Thus

$$p_1 = p_{atm} + \rho g H_1 + \frac{2\sigma}{r_1}$$

$$= 101.3 \times 10^3 + 1000 \times g \times 3 + \frac{2 \times 0.073}{5 \times 10^{-4}}$$

$$= 131.022 \times 10^3 \text{ Nm}^{-2}$$

At a depth of 1 m, the pressure inside the bubble, p_2, is

$$p_2 = p_{atm} + \rho g H_2 + \frac{2\sigma}{r_2}$$

$$= 101.3 \times 10^3 + 1000 \times g \times 1 + \frac{2 \times 0.073}{r_2}$$

$$= 111.11 \times 10^3 + \frac{0.146}{r_2}$$

Since pV is a constant, then

$$p_1 V_1 = p_2 V_2$$

where for a spherical bubble

$$p_1 \frac{4}{3} \pi r_1^3 = p_2 \frac{4}{3} \pi r_2^3$$

That is

$$p_1 r_1^3 = p_2 r_2^3$$

Therefore

$$131.022 \times 10^3 \times (5 \times 10^{-4})^3 = \left(111.11 \times 10^3 + \frac{0.146}{r_2} \right) \times r_2^3$$

The cubic equation can be solved analytically, by trial and error or by assuming that the second term in the brackets is substantially small, reducing the effort required for solution to yield a bubble radius of approximately 0.053 mm.

1.8 Pressure measurement by differential manometer

Determine the pressure difference between two tapping points on a pipe carrying water for a differential manometer reading of 20 cm of mercury. The specific gravity of mercury is 13.6.

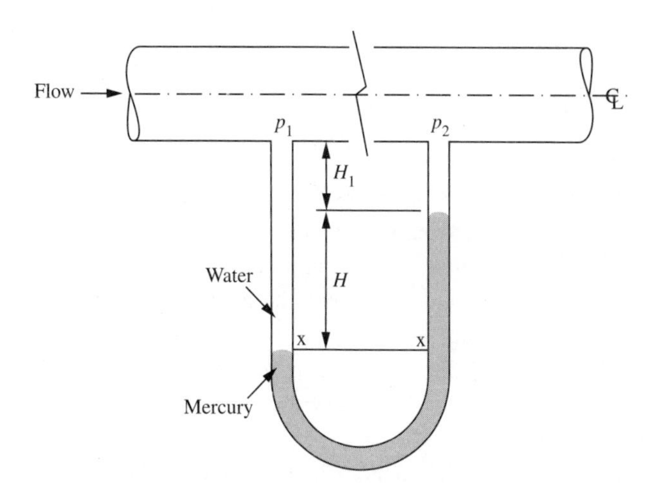

Solution

The differential or U-tube manometer is a device used to measure the difference in pressures between two points and consists of a transparent U-tube, usually made of glass, and contains a manometric fluid such as mercury. It is typically used to measure the pressure drop of moving fluids due to friction along pipes or due to obstacles in pipelines such as flow measuring devices, fittings and changes in geometry. The pressure difference of the process fluid is indicated by the difference in levels of the manometric fluid between the two vertical legs of the U-tube which, at the datum elevation *xx*, are

$$p_1 + \rho g(H_1 + H) = p_2 + \rho g H_1 + \rho_{Hg} g H$$

where ρ_{Hg} is the density of mercury and ρ the density of water.

Rearranging, the differential pressure Δp between the legs is

$$\begin{aligned}\Delta p &= p_1 - p_2 \\ &= \rho g H_1 + \rho_{Hg} g H - \rho g (H_1 + H) \\ &= (\rho_{Hg} - \rho) g H \\ &= (13{,}600 - 1000) \times g \times 0.2 \\ &= 24.721 \times 10^3 \text{ Nm}^{-2}\end{aligned}$$

The differential pressure is 24.7 kNm^{-2}.

Note that the location of the manometer below the pipe, H_1, is not required in the calculation. In practice it is important to allow sufficient length in the legs to prevent the manometric fluid reaching the tapping point on the pipe for high differential pressures. Filled with mercury, differential manometers can typically be used to measure differentials up to about 200 kNm^{-2} or with water to about 20 kNm^{-2}. Where a temperature variation in the process fluid is expected, it is important to allow for density-temperature variation of the manometric fluid, which can affect readings.

In general, the U-tube differential manometer as a pressure-measuring device is largely obsolete. There are many sophisticated methods and pressure-measuring devices now used by industry. But the differential manometer continues to be a useful tool in the laboratory and for testing purposes.

1.9 Pressure measurement by inverted manometer

A laboratory rig is used to examine the frictional losses in small pipes. Determine the pressure drop in a pipe carrying water if a differential head of 40 cm is recorded using an inverted manometer.

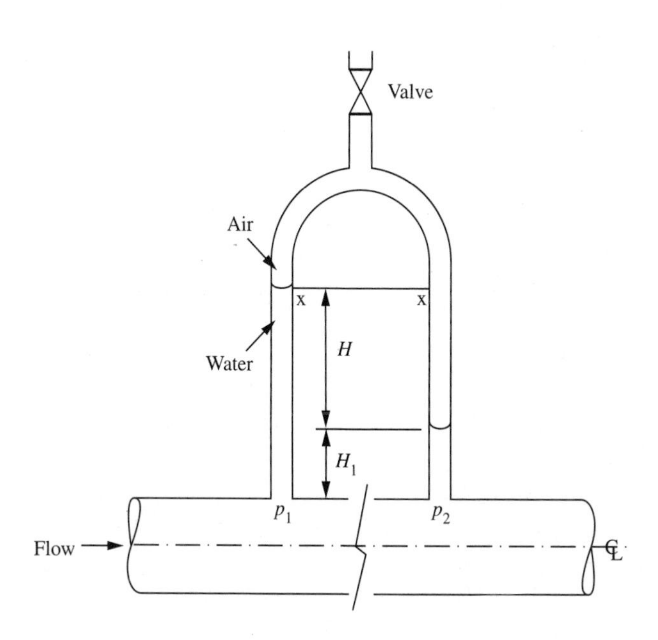

Solution

The inverted manometer avoids the use of a manometric fluid and instead uses the process fluid (water in this case) to measure its own pressure. It consists of an inverted U-tube with a valve into which air or an inert gas can be added or vented. Here, the pressure at the datum elevation xx, in left and right hand legs is

$$p_1 - \rho g(H + H_1) = p_2 - \rho g H_1 - \rho_{air} g H$$

where ρ is the density of water and ρ_{air} is the density of air. Rearranging, the differential pressure Δp is therefore

$$\begin{aligned}\Delta p &= p_1 - p_2 \\ &= \rho g\,(H + H_1) - \rho g H_1 - \rho_{air} g H \\ &= (\rho - \rho_{air}) g H\end{aligned}$$

Since the density of air is in the order of 1000 times less than that of water, it may therefore be reasonably assumed that the differential pressure is approximated to

$$\begin{aligned}\Delta p &\approx \rho g H \\ &\approx 1000 \times g \times 0.4 \\ &= 3924\ \mathrm{Nm}^{-2}\end{aligned}$$

The differential pressure is found to be 3.9 kNm^{-2}. As with the differential manometer, the elevation of the manometer, H_1, is not required in the calculation. In practice, however, it is important to ensure a reasonable position of liquid levels in the legs. This is best achieved by pressurizing the manometer with air or inert gas using the valve, where for high pressures the density may become appreciable and should be taken into consideration. In the case of air, the error in the calculation is unlikely to be greater than 0.5%. In the case illustrated, the density of water corresponds to a temperature of 10°C for which the density of air at atmospheric pressure is 1.2 kgm^{-3}. If this had been taken into account, it would have yielded a differential pressure of 3919 Nm^{-2} or an error of 0.12%. A more significant error is likely to be due to the effects of temperature on density and may affect the result by as much as 1%. Other errors are likely to be caused by defining the top level of the manometric fluid in the vertical leg due to its meniscus. A column-height accuracy of 0.025 mm is, however, generally achievable with the keenest eye reading.

1.10 Pressure measurement by single leg manometer

A mercury-filled single leg manometer is used to measure the pressure drop across a section of plant containing a process fluid of density 700 kgm^{-3}. The pressure drop is maintained by an electrical device which works on an on/off principle using a contact arrangement in a narrow vertical tube of diameter 2 mm while the sump has a diameter of 2 cm. If the pressure drop across the plant is to be increased by 20 kNm^{-2}, determine the quantity of mercury to be removed from the sump if the position of the electrical contact cannot be altered.

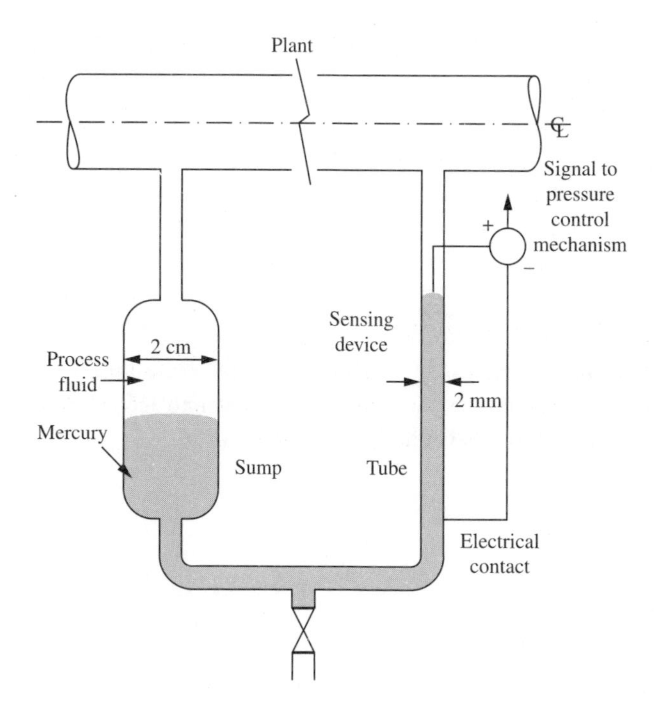

Solution

The single leg manometer uses a sump or reservoir of large cross-section in place of one leg. When a differential pressure is applied, the level in the leg or tube rises due to a displacement from the sump. The ratio of leg to sump area is generally needed for particularly accurate work but is ignored for most purposes since the area of the sump is comparatively larger than that of the leg. The device in this case operates when the level of mercury in the tube falls, breaking the electrical circuit. The pressure control mechanism therefore receives a signal to increase the pressure difference. When the mercury level rises, the opposite occurs. An increase in pressure drop of 20 kNm^{-2} therefore corresponds to an increase in difference in level of mercury of

$$H = \frac{\Delta p}{(\rho_{Hg} - \rho)g}$$

$$= \frac{20 \times 10^3}{(13{,}600 - 700) \times g}$$

$$= 0.158 \text{ m}$$

The volume of mercury to be removed to ensure the contact is still just made is therefore

$$V = \frac{\pi d^2}{4} H$$

$$= \frac{\pi \times 0.02^2}{4} \times 0.158$$

$$= 4.96 \times 10^{-5} \text{ m}^3$$

That is, the volume to be removed is approximately 50 ml. Note that if the displacement of mercury from the sump into the tube is taken into account then this would correspond to a drop in level in the sump, H_s, of

$$H_s = \frac{a}{A} H$$

$$= \left(\frac{d}{d_s}\right)^2 H$$

$$= \left(\frac{0.0002}{0.02}\right)^2 \times 0.158$$

$$= 1.58 \times 10^{-5} \text{ m}$$

This is very small and ignoring it is justified.

1.11 Pressure measurement by inclined leg manometer

An oil-filled inclined leg manometer is used to measure small pressure changes across an air filter in a process vent pipe. If the oil travels a distance of 12 cm along the leg which is inclined at an angle of 20° to the horizontal, determine the gauge pressure across the filter. The density of oil is 800 kgm^{-3}.

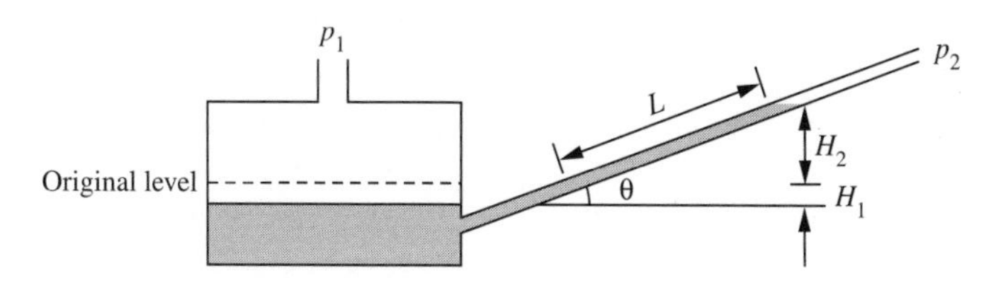

Solution

This instrument is useful for measuring small differential pressures and consists of a sump of manometric fluid (oil) with a leg extended down into it and inclined at some small angle. Applying a differential pressure across the sump and the leg results in a displacement of the manometric fluid into the leg, the distance the manometric liquid travels up along the leg being a measure of differential pressure and is

$$\begin{aligned}\Delta p &= p_1 - p_2 \\ &= \rho g(H_1 + H_2)\end{aligned}$$

If the oil is displaced from the sump up along the leg by a distance L, the corresponding drop in level in the sump, H_1, is therefore

$$H_1 = \frac{aL}{A}$$

Also, the vertical rise of the oil is related to length by the sine of the angle of the inclined leg. That is

$$H_2 = L\sin\theta$$

The differential pressure is therefore

$$\Delta p = \rho g\left(\frac{aL}{A} + L\sin\theta\right)$$

$$= \rho g L\left(\frac{a}{A} + \sin\theta\right)$$

As no details are provided regarding the dimensions of the manometer, the cross-sectional area of the oil sump, A, is therefore assumed to be very much larger than the area of the leg, a. The equation therefore reduces to

$$\Delta p = \rho g L\sin\theta$$

$$= 800 \times g \times 0.12 \times \sin 20°$$

$$= 322 \text{ Nm}^{-2}$$

The differential pressure is found to be 322 Nm^{-2}.

The device is particularly useful for measuring small differential pressures since if the terms inside the brackets are kept small it allows the length along the inclined leg, L, to be appreciable. If, for a given differential pressure, the equivalent movement of manometric liquid up a vertical leg would be h, say, then the ratio of movements L to h

$$\frac{L}{h} = \frac{1}{\dfrac{a}{A} + \sin\theta}$$

can therefore be considered as a magnification ratio.

1.12 Archimedes' principle

A vessel containing a process material with a combined total mass of 100 kg is immersed in water for cooling purposes. Determine the tension in the cable of an overhead crane used to manoeuvre the fully immersed container into its storage position if the bulk density of the vessel is 7930 kgm^{-3}.

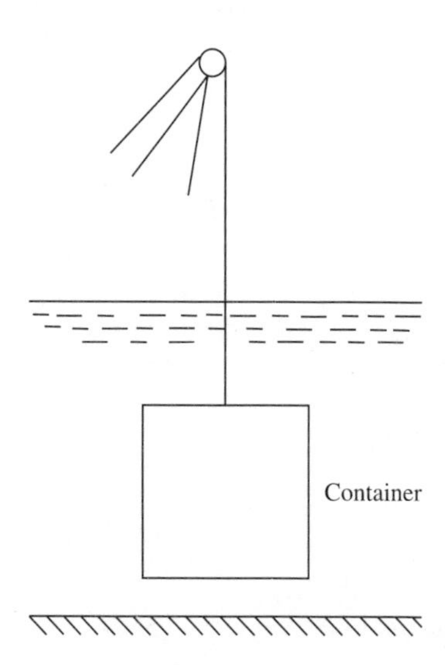

Solution

Consider a body of mass m_c immersed in the liquid such that the net downward force is the difference between the downward and upward forces. That is

$$F = m_c g - mg$$

where m is the mass of water displaced. This is known as Archimedes' principle and states that when a body is partially or totally immersed, there is an upthrust equal to the weight of fluid displaced. For the immersed object, the net downward force is taken by the tension in the cable and can be determined where the mass of the container and water displaced is related to volume by

$$V = \frac{m_c}{\rho_c}$$

$$= \frac{m}{\rho}$$

where ρ_c is the bulk density of the container and ρ is the density of water. Rearranging, the mass of water displaced by the container is therefore

$$m = m_c \frac{\rho}{\rho_c}$$

The tension in the cable is therefore

$$\begin{aligned} F &= m_c g\left(1 - \frac{\rho}{\rho_c}\right) \\ &= 100 \times g \times \left(1 - \frac{1000}{7930}\right) \\ &= 857 \text{ N} \end{aligned}$$

That is, the tension in the cable is 857 N. Note that the tension in the cable when the vessel is lifted out of the water is

$$\begin{aligned} f &= mg \\ &= 100 \times g \\ &= 981 \text{ N} \end{aligned}$$

The buoyancy effect therefore reduces the tension in the cable by 124 N.

1.13 Specific gravity measurement by hydrometer

A hydrometer floats in water with 6 cm of its graduated stem unimmersed, and in oil of SG 0.8 with 4 cm of the stem unimmersed. Determine the length of stem unimmersed when the hydrometer is placed in a liquid of SG 0.9.

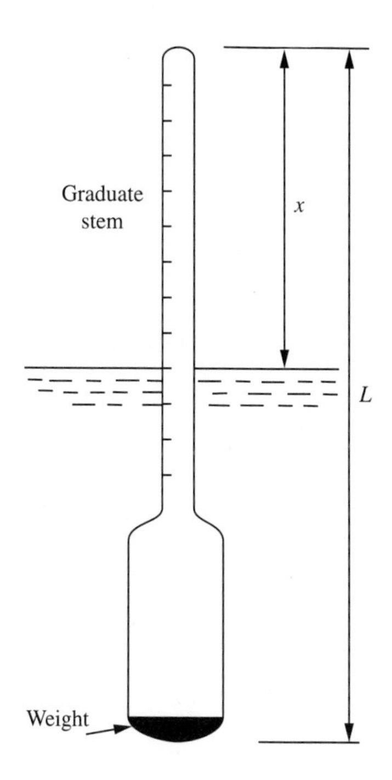

Solution

Hydrometers are simple devices for measuring the density or specific gravity of liquids routinely used in the brewing industry to determine quickly the conversion of sugar to alcohol in fermentation. They consist of a glass tube which have a weighted glass bulb and graduated stem of uniform diameter and float in the liquid being tested. The density or specific gravity (SG) is usually read directly from the graduated stem at the depth to which it sinks. For no net downward force, the vertical downward forces acting on the body are equal to the upthrust. Thus

$$mg = m_h g$$

where m and m_h are the mass of liquid and hydrometer, respectively. Thus, Archimedes' principle for a floating body states that when a body floats, it displaces a weight of fluid equal to its own weight. The displacement by the hydrometer is therefore

$$m_h g = \rho g(L - x)a$$
$$= \rho_o g(L - x_o)a$$

where a is the cross-sectional area of the stem, ρ and ρ_o are the densities of water and oil, and x and x_o are the lengths of stem unimmersed in the respective liquids. Therefore

$$\rho g(L - x)a = \rho_o g(L - x_o)a$$

Rearranging, the length of hydrometer is therefore

$$L = \frac{\rho x - \rho_o x_o}{\rho_o - \rho}$$
$$= \frac{1000 \times 0.06 - 800 \times 0.04}{800 - 1000}$$
$$= 0.14 \text{ m}$$

For the hydrometer immersed in a liquid of SG 0.9 (900 kgm^{-3}), let the length of stem remaining unimmersed be x_L. Therefore

$$1000 \times g \times (0.14 - 0.06) \times a = 900 \times g \times (0.14 - x_L) \times a$$

Solving, x_L is found to be 0.0511 m. That is, the length of stem above the liquid of SG 0.9 is 5.11 cm.

1.14 Transfer of process liquid to a ship

A liquid hydrocarbon mixture of density 950 kgm^{-3} is transferred by pipeline to a ship at a loading terminal. Prior to transfer, the ship has an unloaded displacement of 5000 tonnes and draft of 3 m. Transfer of the hydrocarbon is at a steady rate of 125 m^{3}h^{-1}. If the sea bed is at a depth of 5.5 m, determine the quantity delivered and time taken if the ship requires at least 1 m of clearance between the sea bed and hull to manoeuvre away safely from the loading terminal.

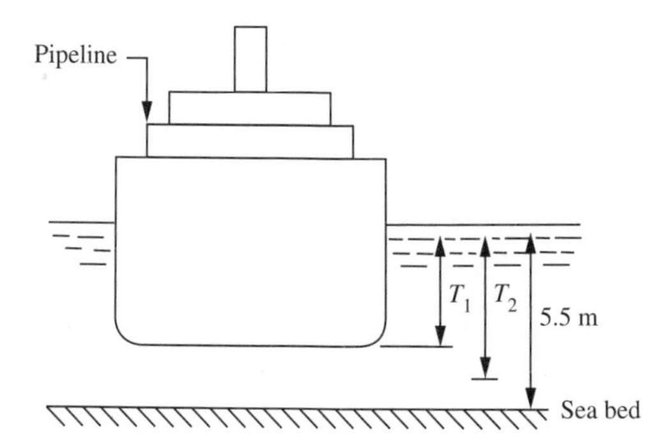

Solution

Applying Archimedes' principle, the ship prior to transfer displaces its own weight of sea water. That is

$$m_s g = mg$$

$$= \rho A T_1 g$$

where m_s and m are the mass of ship and sea water displaced, ρ is the density of sea water, A is the water plane area and T_1 is the depth of the ship below the waterline. After transfer

$$m_s g + m_{hc} g = \rho A T_2 g$$

where m_{hc} is the mass of hydrocarbon mixture. After transfer, the ship is clear from the sea bed by 1 m. Combining these two equations, the mass of hydrocarbon transferred is

$$m_{hc} = m_s\left(\frac{T_2}{T_1} - 1\right)$$

$$= 5 \times 10^6 \times \left(\frac{4.5}{3} - 1\right)$$

$$= 2.5 \times 10^6 \text{ kg}$$

The transfer time is therefore

$$t = \frac{m_{hc}}{\rho_{hc} Q}$$

$$= \frac{2.5 \times 10^6}{950 \times 125}$$

$$= 21.05 \text{ h}$$

That is, a transfer of 2500 tonnes of hydrocarbon mixture is completed in 21.05 hours.

It should be noted that the approach illustrated is rather simplistic. No account is made for the dimensions of the ship in terms of its length and beam nor the variation of the water plane area with depth. The beam is an important dimension in terms of stability where the stability is dependent on the relative position of the ship's centre of gravity and centroid of the displaced volume called the centre of buoyancy. A ship is unstable and will capsize when, for a heel of up to 10°, a line drawn vertically up from the centre of buoyancy is below the centre of gravity — a point known as the metacentre.

The safe transfer of liquids to and from tanks within ships requires a careful sequence of operation. Tidal effects on moored ships and the effects of the liquid free surface in the tanks must also be taken into consideration. It was the British politician Samuel Plimsoll (1824–1898) who was responsible for getting legislation passed to prohibit 'coffinships' — unseaworthy and overloaded ships — being sent to sea. The Merchant Sea Act of 1874 included, amongst other things, enforcement of the painting of lines, originally called Plimsoll marks and now known as load line marks, to indicate the maximum load line which allows for the different densities of the world's seas in summer and winter.

Further problems

(1) Explain what is meant by gauge pressure and absolute pressure.

(2) A hydraulic press has a ram of 10 cm diameter and a plunger of 1 cm diameter. Determine the force required on the plunger to raise a mass of 500 kg on the ram.

Answer: 49.05 N

(3) The reading of a barometer is 75.5 cm of mercury. If the specific gravity of mercury is 13.6, convert this pressure to Newtons per square metre.

Answer: 100,792 Nm^{-2}

(4) A rectangular tank 5 m long by 2 m wide contains water to a depth of 2 m. Determine the intensity of pressure on the base of the tank and the total pressure on the end.

Answer: 19.6 kNm^{-2}, 39 kNm^{-2}

(5) Determine the total pressure on a vertical square sluice, of 1 m square, positioned with its top edge 3 m below the level of water.

Answer: 34.3 kNm^{-2}

(6) A tube is filled with water to a depth of 600 mm and then 450 mm of oil of SG 0.75 is added and allowed to come to rest. Determine the gauge pressure at the common liquid surface and at the base of the tube.

Answer: 3.3 kNm^{-2}, 9.2 kNm^{-2}

(7) Show that when a body is partially or totally immersed in a liquid, there is an upthrust on the body equal to the weight of the liquid displaced.

(8) Show that a floating body displaces a weight of the liquid equal to its own weight.

(9) A U-tube has a left-hand leg with a diameter of 5 cm and a right-hand leg with a diameter of 1 cm and inclined at an angle of 24°. If the manometer fluid is oil with a density of 920 kgm^{-3} and a pressure of 400 Nm^{-2} is applied to the left-hand leg, determine the length by which the oil will have moved along the right-hand leg.

Answer: 9.9 cm

(10) Determine the absolute pressure in an open tank containing crude oil of density 900 kgm^{-3} at a depth of 5 m.

Answer: 145.4 kNm^{-2}

(11) An open storage tank 3 m high contains acetic acid, of density 1060 kgm^{-3}, and is filled to half capacity. Determine the absolute pressure at the bottom of the tank if the vapour space above the acid is maintained at atmospheric pressure.

Answer: 117 kNm^{-2}

(12) A differential manometer containing mercury of SG 13.6 and water indicates a head difference of 30 cm. Determine the pressure difference across the legs.

Answer: 37.1 kNm^{-2}

(13) A U-tube contains water and oil. The oil, of density 800 kgm^{-3}, rests on the surface of the water in the right-hand leg to a depth of 5 cm. If the level of water in the left-hand leg is 10 cm above the level of water in the right-hand leg, determine the pressure difference between the two legs. The density of water is 1000 kgm^{-3}.

Answer: 589 Nm^{-2}

(14) A separator receives continuously an immiscible mixture of solvent and aqueous liquids which is allowed to settle into separate layers. The separator operates with a constant depth of 2.15 m by way of an overflow and underflow arrangement from both layers. The position of the liquid-liquid interface is monitored using a dip leg through which air is gently bubbled. Determine the position of the interface below the surface for a gauge pressure in the dip leg of 20 kNm^{-2}. The densities of the solvent and aqueous phases are 865 kgm^{-3} and 1050 kgm^{-3}, respectively, and the dip leg protrudes to within 5 cm of the bottom of the separator.

Answer: 90 cm

(15) A hydrometer with a mass of 27 g has a bulb of diameter 2 cm and length 8 cm, and a stem of diameter 0.5 cm and length 15 cm. Determine the specific gravity of a liquid if the hydrometer floats with 5 cm of the stem immersed.

Answer: 1.034

(16) Two pressure tapping points, separated by a vertical distance of 12.7 m, are used to measure the crystal content of a solution of sodium sulphate in an evaporator. Determine the density of the solution containing 25% crystals by volume and the differential pressure if the density of the anhydrous sodium sulphate is 2698 kgm^{-3} and the density of saturated sodium sulphate solution is 1270 kgm^{-3}.

Answer: 1627 kgm^{-3}, 203 kNm^{-2}

(17) A vacuum gauge consists of a U-tube containing mercury open to atmosphere. Determine the absolute pressure in the apparatus to which it is attached when the difference in levels of mercury is 60 cm.

Answer: 21.3 kNm^{-2}

(18) Determine the height through which water is elevated by capillarity in a glass tube of internal diameter 3 mm if the hydrostatic pressure is equal to $4\sigma/d$ where σ is the surface tension (0.073 Nm^{-1}) and d is the diameter of the tube.

Answer: 9.9 mm

(19) Explain the effect of surface tension on the readings of gauges of small bore such as piezometer tubes.

(20) A ship has a displacement of 3000 tonnes in sea water. Determine the volume of the ship below the water line if the density of sea water is 1021 kgm^{-3}.

Answer: 2938 m^3

(21) A closed cylindrical steel drum of side length 2 m, outer diameter 1.5 m and wall thickness 8 mm is immersed in a jacket containing water at 20°C (density 998 kgm^{-3}). Determine the net downward and upward forces when the drum is both full of water at 20°C and empty. The density of steel is 7980 kgm^{-3}.

Answer: 5.17 kN, –29.4 kN

(22) An oil/water separator contains water of density 998 kgm^{-3} to a depth of 75 cm above which is oil of density 875 kgm^{-3} to a depth of 75 cm. Determine the total force on the vertical side of the separator if it has a square section 1.5 m broad. If the separator is pressurized by air above the oil, explain how this will affect the answer.

Answer: 16 kN

Continuity, momentum and energy

2

Introduction

With regard to fluids in motion, it is convenient to consider initially an idealized form of fluid flow. In assuming the fluid has no viscosity, it is also deemed to have no frictional resistance either within the fluid or between the fluid and pipe walls. Inviscid fluids in motion therefore do not support shear stresses although normal pressure forces still apply.

There are three basic conservation concepts evoked in solving problems involving fluids in motion. The conservation of mass was first considered by Leonardo da Vinci (1452–1519) in 1502 with respect to the flow within a river. Applied to the flow through a pipe the basic premise is that mass is conserved. Assuming no loss from or accumulation within the pipe, the flow into the pipe is equal to the flow out and can be proved mathematically by applying a mass balance over the pipe section. The flow of incompressible fluids at a steady rate is therefore the simplest form of the continuity equation and may be readily applied to liquids.

The conservation of momentum is Newton's second law applied to fluids in motion, and was first considered by the Swiss mathematician Leonhard Euler (1707–1783) in 1750. Again, by considering inviscid fluid flow under steady flow conditions, calculations are greatly simplified. This approach is often adequate for most engineering purposes.

The conservation of energy was first considered by the Swiss scientist Daniel Bernoulli (1700–1782) in 1738 to describe the conservation of mechanical energy of a moving fluid in a system. The basic premise is that the total energy of the fluid flowing in a pipe must be conserved. An energy balance on the moving fluid across the pipe takes into account the reversible pressure-volume, kinetic and potential energy forms, and is greatly simplified by considering steady, inviscid and incompressible fluid flow.

2.1 Flow in branched pipes

Water flows through a pipe section with an inside diameter of 150 mm at a rate of 0.02 m^3s^{-1}. The pipe branches into two smaller diameter pipes, one with an inside diameter of 50 mm and the other with an inside diameter of 100 mm. If the average velocity in the 50 mm pipe is 3 ms^{-1}, determine the velocities and flows in all three pipe sections.

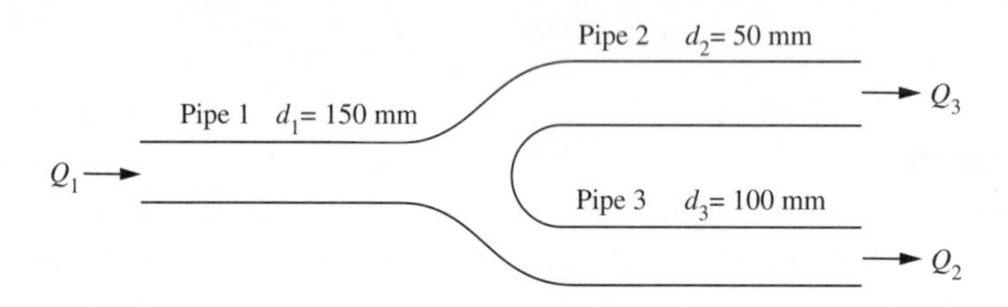

Solution

The continuity equation is effectively a mathematical statement describing the conservation of mass of a flowing fluid where the mass flow into a pipe section is equal to the mass flow out. That is

$$\rho_1 a_1 v_1 = \rho_2 a_2 v_2$$

For an incompressible fluid in which the density does not change, the volumetric flow is therefore

$$a_1 v_1 = a_2 v_2$$

For the branched pipe system in which there is no loss or accumulation of the incompressible process fluid (water), the flow through the 150 mm diameter pipe (Pipe 1) is equal to the sum of flows in the 50 mm (Pipe 2) and 100 mm diameter pipes (Pipe 3). That is

$$Q_1 = Q_2 + Q_3$$

$$= \frac{\pi d_2^2}{4} v_2 + \frac{\pi d_3^2}{4} v_3$$

Rearranging, the velocity in the 100 mm diameter pipe is therefore

$$v_3 = \frac{\frac{4Q_1}{\pi} - d_2^2 v_2}{d_3^2}$$

$$= \frac{\frac{4 \times 0.02}{\pi} - 0.05^2 \times 3}{0.1^2}$$

$$= 1.8 \text{ ms}^{-1}$$

This corresponds to a flow of

$$Q_3 = \frac{\pi d_3^2}{4} v_3$$

$$= \frac{\pi \times 0.1^2}{4} \times 1.8$$

$$= 0.014 \text{ m}^3\text{s}^{-1}$$

Similarly, the velocity and flow can be found for the other two pipes and are given below.

	Pipe 1	Pipe 2	Pipe 3
Diameter, mm	150	50	100
Velocity, ms^{-1}	1.13	3.00	1.80
Flowrate, m^3s^{-1}	0.020	0.006	0.014

2.2 Forces on a U-bend

A horizontal pipe has a 180° U-bend with a uniform inside diameter of 200 mm and carries a liquid petroleum fraction of density 900 kgm^{-3} at a rate of 150 m^{3}h^{-1}. Determine the force exerted by the liquid on the bend if the gauge pressure upstream and downstream of the bend are 100 kNm^{-2} and 80 kNm^{-2}, respectively.

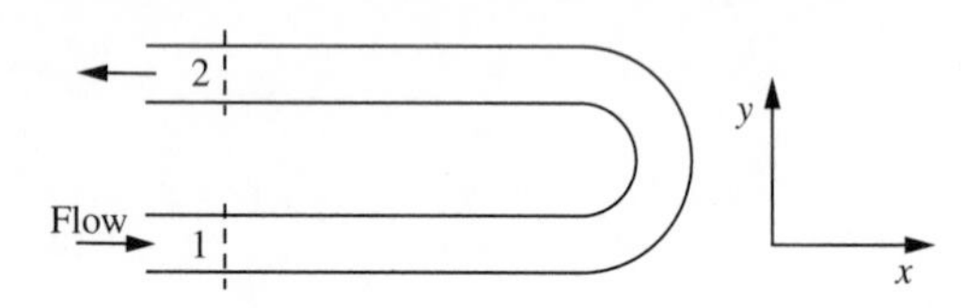

Solution

The thrust exerted by the flowing liquid on the horizontal bend is resolved in both the x- and y-directions. Assuming that the gauge pressures of the liquid are distributed uniformly in the U-bend, then resolving the force in the x-direction gives

$$F_x = p_1 a_1 \cos\theta_1 - p_2 a_2 \cos\theta_2 + \rho Q(v_2 \cos\theta_2 - v_1 \cos\theta_1)$$

and in the y-direction

$$F_y = p_1 a_1 \sin\theta_1 + p_2 a_2 \sin\theta_2 - \rho Q(v_2 \sin\theta_2 + v_1 \sin\theta_1)$$

The respective upstream and downstream pressure forces are

$$p_1 a_1 = 1 \times 10^5 \times \frac{\pi \times 0.2^2}{4}$$

$$= 3141 \text{ N}$$

and

$$p_2 a_2 = 8 \times 10^4 \times \frac{\pi \times 0.2^2}{4}$$

$$= 2513 \text{ N}$$

For the uniform cross-section, the average velocity remains constant. That is

$$\begin{aligned} v_1 &= v_2 \\ &= \frac{4Q}{\pi d^2} \\ &= \frac{4 \times \dfrac{150}{3600}}{\pi \times 0.2^2} \\ &= 1.33 \text{ ms}^{-1} \end{aligned}$$

The momentum fluxes are therefore

$$\begin{aligned} \rho Q v_1 &= \rho Q v_2 \\ &= 900 \times \frac{150}{3600} \times 1.33 \\ &= 49.9 \text{ N} \end{aligned}$$

For the liquid entering the 180° bend the angle θ_1 is 0° and for the liquid leaving θ_2 is 180°. The resolved force in the x-direction is therefore

$$\begin{aligned} F_x &= 3141 \times \cos 0° - 2513 \times \cos 180° + 49.9 \times (\cos 180° - \cos 0°) \\ &= 5554 \text{ N} \end{aligned}$$

Since $\sin 0°$ and $\sin 180°$ are equal to zero, the force in the y-direction is

$$F_y = 0$$

Although not taken into consideration here, the reaction in the vertical direction F_z can also be included where the downward forces are due to the weight of the bend and the fluid contained within it.

2.3 Pressure rise by valve closure

A valve at the end of a water pipeline of 50 mm inside diameter and length 500 m is closed in 1 second giving rise to a uniform reduction in flow. Determine the average pressure rise at the valve if the average velocity of the water in the pipeline before valve closure had been 1.7 ms^{-1}.

Solution

When a liquid flowing along a pipeline is suddenly brought to rest by the closure of a valve or any other obstruction, there will be a large rise in pressure due to the loss of momentum causing a pressure wave to be transmitted along the pipe. The corresponding force on the valve is therefore

$$F = m\frac{v}{t}$$

where v/t is the deceleration of the liquid and the mass of water in the pipeline is

$$m = \rho a L$$

Thus

$$F = \rho a L \frac{v}{t}$$

$$= 1000 \times \frac{\pi \times 0.05^2}{4} \times 500 \times \frac{1.7}{1}$$

$$= 1669 \text{ N}$$

corresponding to a pressure on the valve of

$$p = \frac{F}{a}$$

$$= \frac{4F}{\pi d^2}$$

$$= \frac{4 \times 1669}{\pi \times 0.05^2}$$

$$= 850.015 \times 10^3 \text{ kNm}^{-2}$$

The average pressure on the valve on closure is found to be 850 kNm^{-2}. Serious and damaging effects due to sudden valve closure can occur, however, when the flow is retarded at such a rate that a pressure wave is transmitted back along the pipeline. The maximum (or critical) time in which the water can be brought to rest producing a maximum or peak pressure is

$$t = \frac{L}{c}$$

where c is the velocity of sound transmission through the water. With no resistance at the entrance to the pipeline, the excess pressure is relieved. The pressure wave then travels back along the pipeline reaching the closed valve at a time $2L/c$ later. (The period of $2L/c$ is known as the pipe period.) In practice, closures below values of $2L/c$ are classed as instantaneous. In this problem, the critical time corresponds to 0.67 seconds for a transmission velocity of 1480 ms^{-1} and is below the 1.0 second given. The peak pressure can be significantly greater than the average pressure on valve closure with the pressure wave being transmitted up and down the pipeline until its energy is eventually dissipated. It is therefore important to design piping systems within acceptable design limits. Accumulators (air chambers or surge tanks) or pressure relief valves located near the valves can prevent potential problems.

The peak pressure resulting from valve closures faster than the pipe period can be calculated (in head form) from

$$H = \frac{vc}{g}$$

This basic equation, developed by the Russian scientist N. Joukowsky in 1898, implies that a change in flow directly causes a change in pressure, and vice versa. The velocity of sound transmission, c, is however variable and is dependent upon the physical properties of the pipe and the liquid being conveyed. The presence of entrained gas bubbles markedly decreases the effective velocity of sound in the liquid. In this case, the peak head is

$$H = \frac{1.7 \times 1480}{g}$$

$$= 256.5 \text{ m}$$

which corresponds to a peak pressure of 2516 kNm^{-2}. However, the Joukowsky equation neglects to consider the possible rise due to the reduction in frictional pressure losses that occur as the fluid is brought to rest. It also does not consider the pressure in the liquid that may exist prior to valve closure — all of which may well be in excess of that which can be physically withstood by the pipe.

2.4 The Bernoulli equation

An open tank of water has a pipeline of uniform diameter leading from it as shown below. Neglecting all frictional effects, determine the velocity of water in the pipe and the pressure at points A, B and C.

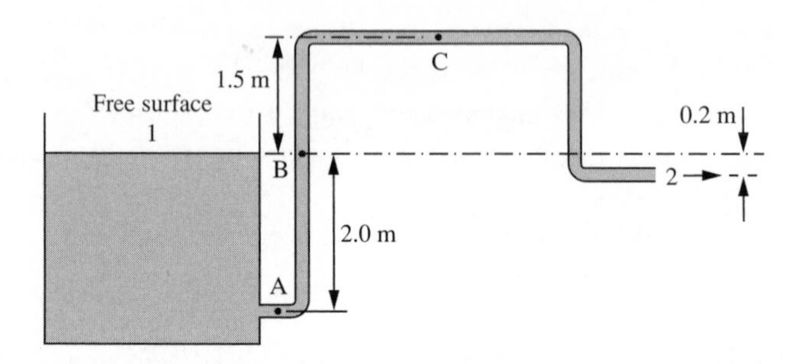

Solution

The Bernoulli equation (named after Daniel Bernoulli) is

$$\frac{p_1}{\rho g} + \frac{v_1^2}{2g} + z_1 = \frac{p_2}{\rho g} + \frac{v_2^2}{2g} + z_2$$

The first, second and third terms of the equation are known as the pressure head, velocity head and static head terms respectively, each of which has the fundamental dimensions of length. This is an important equation for the analysis of fluid flow in which thermodynamic occurrences are not important. It is derived for an incompressible fluid without viscosity. These assumptions give results of acceptable accuracy for liquids of low viscosity and for gases flowing at subsonic speeds when changes in pressure are small.

To determine the velocity in the pipe, the Bernoulli equation is applied between the free surface (point 1) and the end of the pipe (point 2) which are both exposed to atmospheric pressure. That is

$$p_1 = p_2 = p_{atm}$$

The tank is presumed to be of sufficient capacity that the velocity of the water at the free surface is negligible. That is

$$v_1 \approx 0$$

Therefore

$$v_2 = \sqrt{2g(z_1 - z_2)}$$

$$= \sqrt{2g \times 0.2}$$

$$= 1.98 \text{ ms}^{-1}$$

The average velocity is the same at all points along the pipeline. That is

$$v_2 = v_A = v_B = v_C$$

The pressure at A is therefore

$$p_A = \rho g \left(z_1 - z_A - \frac{v_A^2}{2g} \right)$$

$$= 1000g \times \left(2 - \frac{1.98^2}{2g} \right)$$

$$= 17{,}658 \text{ Nm}^{-2}$$

The pressure at B is

$$p_B = \rho g \left(z_1 - z_B - \frac{v_B^2}{2g} \right)$$

$$= 1000g \times \left(0 - \frac{1.98^2}{2g} \right)$$

$$= -1962 \text{ Nm}^{-2}$$

Finally, the pressure at C is

$$p_C = \rho g \left(z_1 - z_C - \frac{v_C^2}{2g} \right)$$

$$= 1000g \times \left(-1.5 - \frac{1.98^2}{2g} \right)$$

$$= -16{,}677 \text{ Nm}^{-2}$$

The average velocity in the pipeline is 1.98 ms^{-1} and the pressures at points A, B and C are 17.658 kNm^{-2}, -1.962 kNm^{-2} and -16.677 kNm^{-2}, respectively.

2.5 Pressure drop due to enlargements

Water flows through a pipe with an inside diameter of 5 cm at a rate of 10 m^3h^{-1} and expands into a pipe of inside diameter 10 cm. Determine the pressure drop across the pipe enlargement.

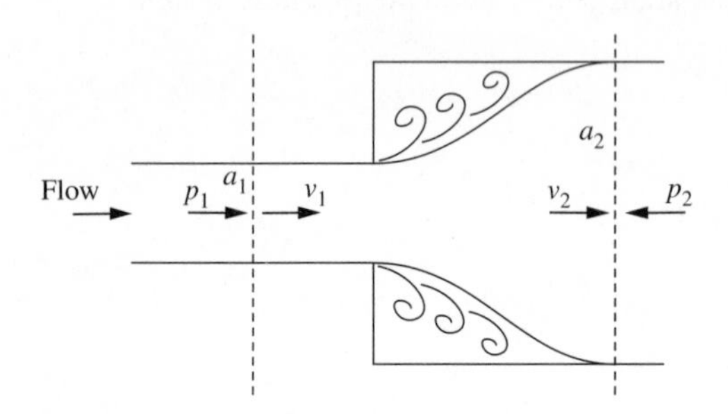

Solution

If a pipe suddenly enlarges, eddies form at the corners and there is a permanent and irreversible energy loss. A momentum balance across the enlargement gives

$$p_1 a_2 + \rho Q v_1 = p_2 a_2 + \rho Q v_2$$

Rearranging, the pressure drop is therefore

$$p_2 - p_1 = \frac{\rho Q (v_1 - v_2)}{a_2}$$

where the average velocity in the smaller pipe is

$$v_1 = \frac{4Q}{\pi d^2}$$

$$= \frac{4 \times \frac{10}{3600}}{\pi \times 0.05^2}$$

$$= 1.41 \text{ ms}^{-1}$$

and in the larger pipe is

$$v_2 = \left(\frac{d_1}{d_2}\right)^2 v_1$$

$$= \left(\frac{0.05}{0.1}\right)^2 \times 1.41$$

$$= 0.352 \text{ ms}^{-1}$$

The pressure drop is therefore

$$p_2 - p_1 = \frac{1000 \times \dfrac{10}{3600} \times (1.41 - 0.352)}{\dfrac{\pi \times 0.1^2}{4}}$$

$$= 374 \text{ Nm}^{-2}$$

Applying the Bernoulli equation over the section, the head loss is

$$H_L = \frac{v_1^2 - v_2^2}{2g} + \frac{p_1 - p_2}{\rho g}$$

$$= \frac{v_1^2 - v_2^2}{2g} - \frac{\rho Q(v_1 - v_2)}{a_2 \rho g}$$

$$= \frac{v_1^2 - v_2^2}{2g} - \frac{2v_2(v_1 - v_2)}{2g}$$

which reduces to

$$H_L = \frac{(v_1 - v_2)^2}{2g}$$

$$= \frac{v_1^2}{2g}\left(1 - \frac{v_2}{v_1}\right)^2$$

From continuity for an incompressible fluid

$$a_1 v_1 = a_2 v_2$$

Then

$$H_L = \frac{v_1^2}{2g}\left(1 - \frac{a_1}{a_2}\right)^2$$

or in terms of diameter for the circular pipe

$$H_L = \frac{v_1^2}{2g}\left(1 - \left(\frac{d_1}{d_2}\right)^2\right)^2$$

Therefore

$$H_L = \frac{1.41^2}{2g} \times \left(1 - \left(\frac{0.05}{0.1}\right)^2\right)^2$$

$$= 0.057 \text{ m}$$

The pressure drop is therefore

$$\Delta p_f = \rho g H_L$$

$$= 1000 \times g \times 0.057$$

$$= 559 \text{ Nm}^{-2}$$

The pressure drop is 559 Nm^{-2}. Note that for a considerable enlargement where $a_2 >> a_1$ the head loss tends to

$$H_L = \frac{v_1^2}{2g}$$

That is, the head loss due to an enlargement is equal to one velocity head based on the velocity in the smaller pipe. This is often referred to as the pipe exit head loss. Note that although there is a loss of energy (or head) there may not necessarily be a drop in fluid pressure because the increase in cross-section causes a reduction in velocity and an increase in pressure.

2.6 Pipe entrance head loss

Derive an expression for the entrance loss in head form for a fluid flowing through a pipe abruptly entering a pipe of smaller diameter.

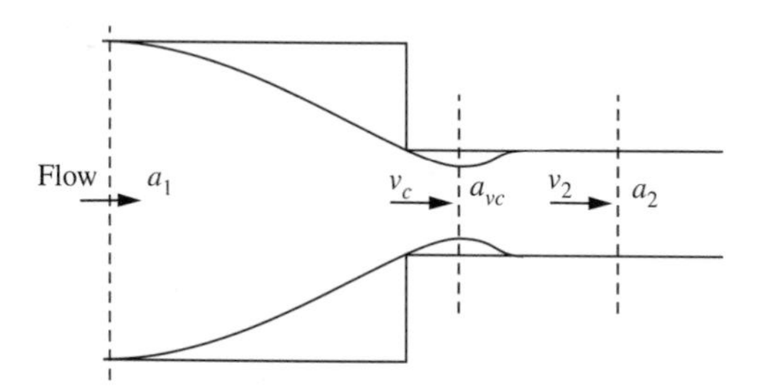

Solution

The permanent and irreversible loss of head due to a sudden contraction is not due to the sudden contraction itself, but due to the sudden enlargement following the contraction. Consider, therefore, a pipe of area a_1 which reduces to area a_2. The fluid flowing into the narrow pipe is further contracted forming a *vena contracta*. At this point the area a_{vc} is related to the smaller pipe area by a coefficient of contraction as

$$a_{vc} = C_c a_2$$

Beyond the *vena contracta*, the fluid expands and fills the pipe. The head loss due to this expansion is

$$H_L = \frac{(v_{vc} - v_2)^2}{2g}$$

and is known as the Carnot-Borda equation after the French mathematicians Lazare Nicolas Marguerite Carnot (1753–1823) and Jean Charles Borda (1733–1799). From continuity

$$\begin{aligned} a_2 v_2 &= a_{vc} v_{vc} \\ &= C_c a_2 v_{vc} \end{aligned}$$

Therefore

$$v_{vc} = \frac{v_2}{C_c}$$

Then

$$H_L = \frac{v_2^2}{2g}\left(\frac{1}{C_c^2} - 1\right)^2$$

$$= k\frac{v_2^2}{2g}$$

The constant k is found by experiment. For a sudden contraction, the head loss is close to

$$H_L = 0.5\frac{v_2^2}{2g}$$

and is usually referred to as the entrance head loss to a pipe. Experimental values are

a_2/a_1	0	0.2	0.4	0.6	0.8	1.0
k	0.5	0.45	0.36	0.21	0.07	0

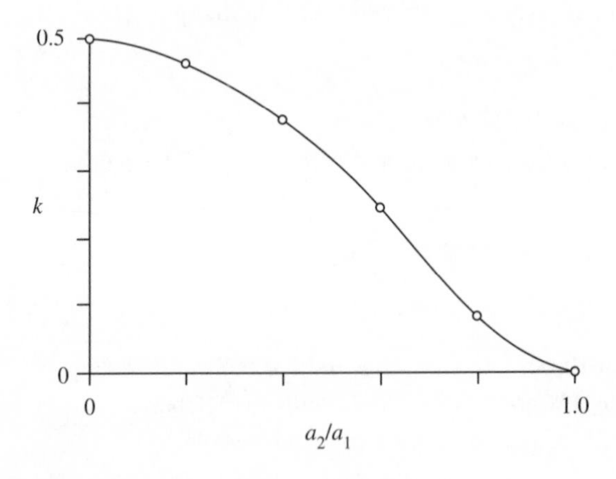

2.7 Force on a pipe reducer

Water flows through a pipe of inside diameter 200 mm at a rate of 100 m^{3}h^{-1}. If the flow abruptly enters a section reducing the pipe diameter to 150 mm, for which the head loss is 0.2 velocity heads based on the smaller pipe, determine the force required to hold the section in position. Upstream of the reducer, the gauge pressure is 80 kNm^{-2}.

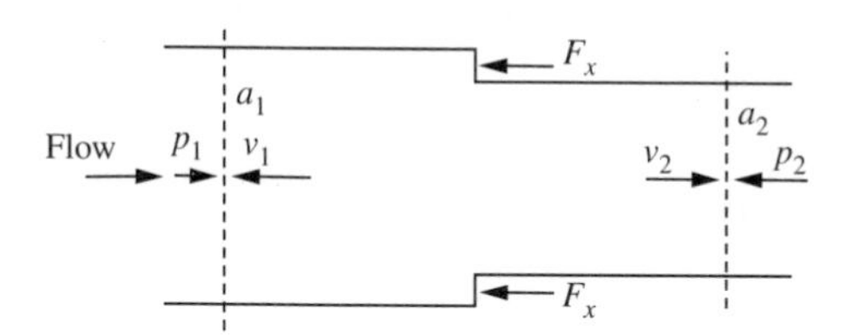

Solution

The velocities in the larger and smaller pipe are

$$v_1 = \frac{4Q}{\pi d_1^2}$$

$$= \frac{4 \times \dfrac{100}{3600}}{\pi \times 0.2^2}$$

$$= 0.884 \text{ ms}^{-1}$$

and

$$v_2 = \frac{4Q}{\pi d_1^2}$$

$$= \frac{4 \times \dfrac{100}{3600}}{\pi \times 0.15^2}$$

$$= 1.57 \text{ ms}^{-1}$$

The head loss at the reducer is based on the velocity in the smaller pipe as

$$H_L = 0.2\frac{v_2^2}{2g}$$

$$= 0.2 \times \frac{1.57^2}{2g}$$

$$= 0.025 \text{ m}$$

The pressure in the 200 mm diameter pipe is 80 kNm^{-2}. The pressure in the 150 mm diameter pipe is found by applying the Bernoulli equation

$$\frac{p_1}{\rho g} + \frac{v_1^2}{2g} = \frac{p_2}{\rho g} + \frac{v_2^2}{2g} + H_L$$

Rearranging

$$p_2 = p_1 + \frac{\rho}{2}(v_1^2 - v_2^2) - \rho g H_L$$

$$= 80 \times 10^3 + \frac{1000}{2} \times (0.884^2 - 1.57^2) - 1000 \times g \times 0.025$$

$$= 78{,}913 \text{ Nm}^{-2}$$

The upstream and downstream pressure forces are therefore

$$p_1 a_1 = 80 \times 10^3 \times \frac{\pi \times 0.2^2}{4}$$

$$= 2513 \text{ N}$$

and

$$p_2 a_2 = 78{,}913 \times \frac{\pi \times 0.15^2}{4}$$

$$= 1394 \text{ N}$$

The force in the x-direction is therefore

$$F_x = \rho Q(v_2 - v_1) - p_1 a_1 + p_2 a_2$$

$$= 1000 \times \frac{100}{3600} \times (1.57 - 0.884) - 2513 + 1394$$

$$= -1100 \text{ N}$$

A force of 1.1 kNm^{-2} in the opposite direction to flow is required to hold the reducing section in position.

2.8 Vortex motion

Derive an expression for the variation of total head across the streamlines of a rotating liquid.

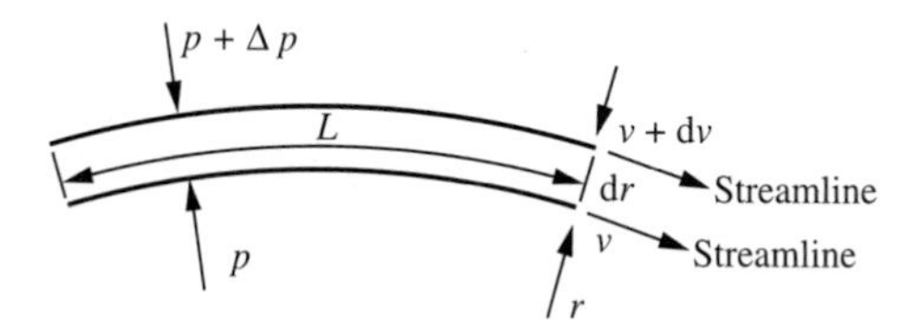

Solution

Consider an element of liquid of length L and width dr between two horizontal streamlines of radii r and r +dr and which have corresponding velocities v and v +dv. The difference in radial force is equal to the centrifugal force. That is

$$\Delta pL = \frac{\rho L v^2 \mathrm{d}r}{r}$$

from which the pressure head is therefore deduced to be

$$\frac{\Delta p}{\rho g} = \frac{v^2 \mathrm{d}r}{gr}$$

The radial rate of change of pressure head is therefore

$$\frac{\mathrm{d}\left(\dfrac{\Delta p}{\rho g}\right)}{\mathrm{d}r} = \frac{v^2}{gr}$$

while the radial change of velocity head is

$$\frac{\mathrm{d}\left(\dfrac{v^2}{2g}\right)}{\mathrm{d}r} = \frac{(v + \mathrm{d}v)^2 - v^2}{2g\mathrm{d}r}$$

$$\approx \frac{v}{g}\frac{\mathrm{d}v}{\mathrm{d}r}$$

The rate of change of total head with radius is therefore

$$\frac{\mathrm{d}H}{\mathrm{d}r} = \frac{v^2}{gr} + \frac{v}{g}\frac{\mathrm{d}v}{\mathrm{d}r}$$

$$= \frac{v}{g}\left(\frac{v}{r} + \frac{\mathrm{d}v}{\mathrm{d}r}\right)$$

This is an important result based on a horizontal moving fluid and can be used to determine the variation of head (or pressure) with radius for both forced and free vortex motion. In a free vortex, the fluid is allowed to rotate freely such as in the case of a whirlwind, flow round a sharp bend or drainage from a plughole. There is a constant total head across the streamline. Thus

$$\mathrm{d}H = 0$$

such that vr is a constant. For free vortex flow it can be shown that the velocity increases and pressure decreases towards the centre.

In a forced vortex in which a fluid is rotated or stirred by mechanical means, the tangential velocity is directly proportional to the streamline radius as

$$v = \omega r$$

where ω is the angular velocity. For forced vortex flow it can be shown that the free surface is a paraboloid.

2.9 Forced and free vortices

An impeller of diameter 50 cm rotating at 60 rpm about the vertical axis inside a large vessel produces a circular vortex motion in the liquid. Inside the impeller region the motion produces a forced vortex and a free vortex outside the impeller with the velocity of the forced and free vortices being assumed equal at the impeller edge. Determine the level of the free surface at a radius equal to the impeller and a considerable distance from the impeller shaft above the liquid surface depression.

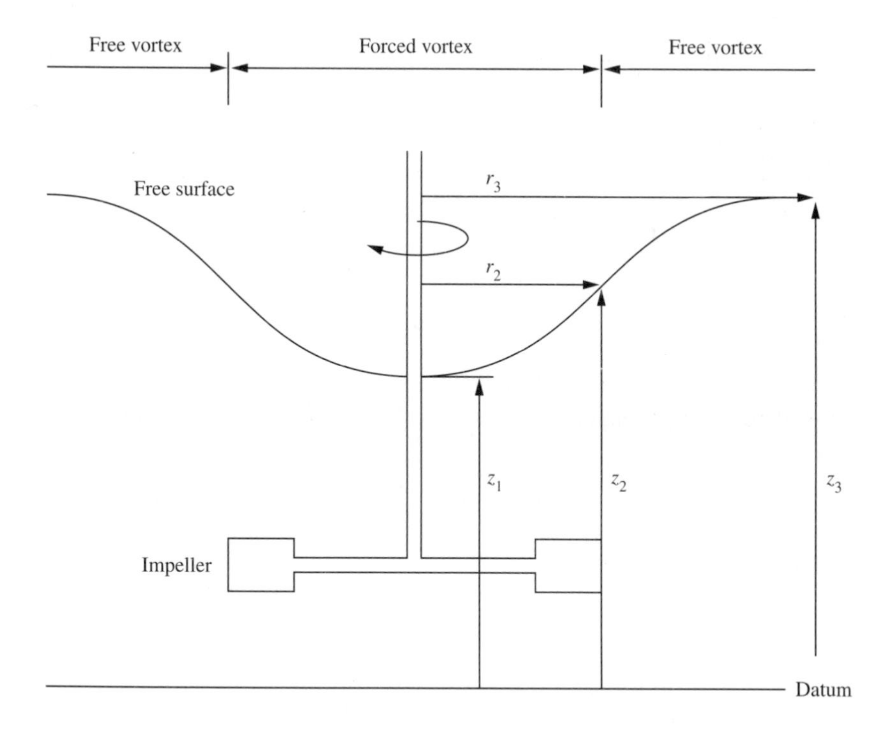

Solution

For two-dimensional flow, the rate of change of total head H with radius r is given by

$$\frac{\mathrm{d}H}{\mathrm{d}r} = \frac{v}{g}\left(\frac{v}{r} + \frac{\mathrm{d}v}{\mathrm{d}r}\right)$$

(see Problem 2.8, page 51)

For the forced vortex, the tangential velocity is related to angular velocity by

$$v = \omega r$$

and thus

$$\mathrm{d}v = \omega \mathrm{d}r$$

Therefore

$$\frac{\mathrm{d}H}{\mathrm{d}r} = \frac{2\omega^2 r}{g}$$

Integrating, the difference in total head between two streamlines of radii r_2 and r_1 is

$$\int_{H_1}^{H_2} \mathrm{d}H = \frac{2\omega^2}{g} \int_{r_1}^{r_2} r \mathrm{d}r$$

to give

$$H_2 - H_1 = \frac{\omega^2 (r_2^2 - r_1^2)}{g}$$

From the Bernoulli equation applied at the free surface ($p_1 = p_2$), the total head is

$$H_2 - H_1 = \frac{v_2^2 - v_1^2}{2g} + z_2 - z_1$$

$$= \frac{\omega^2 (r_2^2 - r_1^2)}{2g} + z_2 - z_1$$

Combining the equations for total head gives

$$z_2 - z_1 = \frac{\omega^2 (r_2^2 - r_1^2)}{2g}$$

Since at the centre of the paraboloid $r_1 = 0$, the elevation of the free surface z_2 at the radius of the impeller r_2 is therefore

$$z_2 - z_1 = \frac{(\omega r_2)^2}{2g}$$

$$= \frac{(2\pi N r_2)^2}{2g}$$

$$= \frac{\left(2 \times \pi \times \frac{60}{60} \times 0.25\right)^2}{2g}$$

$$= 0.126 \text{ m}$$

For the free vortex

$$\mathrm{d}H = 0$$

Therefore

$$\frac{\mathrm{d}v}{v} + \frac{\mathrm{d}r}{r} = 0$$

Integrating

$$\log_e \frac{v_2}{v_1} + \log_e \frac{r_2}{r_1} = 0$$

then for the free surface

$$c = vr$$

where c is a constant. At the edge of the impeller (r_2) this is

$$c = v_2 r_2$$

$$= \omega r_2^2$$

$$= 2\pi N r_2^2$$

$$= 2 \times \pi \times \frac{60}{60} \times 0.25^2$$

$$= 0.393$$

The tangential velocity at the edge of the impeller is

$$v_2 = \frac{c}{r_2}$$

and at a distant point from the impeller is

$$v_3 = \frac{c}{r_3}$$

Applying the Bernoulli equation at the free surface ($p_2 = p_3$) then

$$z_3 - z_2 = \frac{v_2^2 - v_3^2}{2g}$$

$$= \frac{\left(\frac{c}{r_2}\right)^2 - \left(\frac{c}{r_3}\right)^2}{2g}$$

Since the radius r_3 is large then

$$z_3 - z_2 = \frac{c^2}{2gr_2^2}$$

$$= \frac{0.393^2}{2 \times g \times 0.25^2}$$

$$= 0.126 \text{ m}$$

The total depression in the rotating liquid is therefore

$$z_3 - z_1 = (z_3 - z_2) + (z_2 - z_1)$$

$$= 0.126 \text{ m} + 0.126 \text{ m}$$

$$= 0.252 \text{ m}$$

The total depression is found to be 25.2 cm.

Further problems

(1) Water flows upwards through a pipe which tapers from a diameter of 200 mm to 150 mm over a distance of 1 m. Neglecting friction, determine the rate of flow if the gauge pressure at the 200 mm section is 200 kNm^{-2} and at the 150 mm section is 150 kNm^{-2}. Water has a density of 1000 kgm^{-3}.

Answer: 0.192 m^3s^{-1}

(2) Show that for a liquid freely discharging vertically downwards from the end of a pipe, the cross-sectional area of the *vena contracta*, a_2, to pipe area, a_1, separated by a distance z is

$$\frac{2gz}{Q^2} = \frac{1}{a_2^2} - \frac{1}{a_1^2}$$

(3) A pipe of inside diameter 100 mm is suddenly enlarged to a diameter of 200 mm. Determine the loss of head due to this enlargement for a rate of flow of 0.05 m^3s^{-1}.

Answer: 1.16 m

(4) Water is discharged from a tank through an external cylindrical mouthpiece with an area of 100 cm^2 under a pressure of 30 kNm^{-2}. Determine the rate of discharge if the coefficient of contraction is 0.64.

Answer: 0.0645 m^3s^{-1}

(5) Water is added to a process vessel in the form of a jet and directed perpendicularly against a flat plate. If the diameter of the jet is 25 mm and the jet velocity is 10 ms^{-1}, determine the power of the jet and the magnitude of the force acting on the plate.

Answer: 491 W, 49.1 N

(6) A process liquid of density 1039 kgm^{-3} is fed continuously into a vessel as a jet at a rate of 12 m^3h^{-1}. If the jet, which has a diameter of 25 mm, impinges on a flat surface at an angle of 60° to the jet, determine the force on the plate.

Answer: 20.4 N

(7) A jet of water 50 mm in diameter with a velocity of 10 ms^{-1} strikes a series of flat plates normally. If the plates are moving in the same direction as the jet with a velocity of 7 ms^{-1}, determine the pressure on the plates and the work done.

Answer: 30 kNm^{-2}, 412 W

(8) Show that the efficiency η of a simple water wheel consisting of flat plates attached radially around the circumference in which a fluid impinges tangentially is

$$\eta = \frac{2(v_1 - v_2)v_2}{v_1^2}$$

where v_1 is the velocity of the jet and v_2 is the velocity of the plates.

(9) Determine the efficiency of the plates in Further Problem (7).

Answer: 42%

(10) Show that the maximum efficiency of the water wheel described in Further Problem (8) in which a jet impinges normally on its flat vanes is 50%.

(11) Water is discharged through a horizontal nozzle at a rate of 25 litres per second. If the nozzle converges from a diameter of 50 mm to 25 mm and the water is discharged to atmosphere, determine the pressure at the inlet to the nozzle and the force required to hold the nozzle in position.

Answer: 150 kNm^{-2}, 133 N

(12) Two horizontal pipes with inside diameters of 4 cm are connected by a smooth horizontal 90° elbow. Determine the magnitude and direction of the horizontal component of the force which is required to hold the elbow at a flowrate of 25 litres per second and at a gauge pressure of 3 atmospheres. Air at atmospheric pressure surrounds the pipes.

Answer: 1063 N, 45°

(13) A pipeline of inside diameter 30 cm carries crude oil of density 920 kgm^{-3} at a rate of 500 m^3h^{-1}. Determine the force on a 45° horizontal elbow if the pressure in the elbow is constant at 80 kNm^{-2}.

Answer: 4122 N, 22.5°

(14) A 200 mm inside diameter pipe carries a process liquid of density 1017 kgm^{-3} at a rate of 200 m^3h^{-1}. Determine the magnitude and direction of the force acting on a 90° elbow due to momentum change only.

Answer: 141 N, 45°

(15) A pipeline with a diameter of 90 cm carries water with an average velocity of 3 ms^{-1}. Determine the magnitude and direction of the force acting on a 90° bend due to momentum change.

Answer: 8097 N, 45°

(16) Acetic acid with a density of 1070 kgm^{-3} flows along a pipeline at a rate of 54 m^3h^{-1}. The pipeline has an inside diameter of 100 mm and rises to an elevation of 5 m. Determine the kinetic energy per unit volume of the acid and the pressure head at the elevated point if the gauge pressure at the lower elevation is 125 kNm^{-2}.

Answer: 1951 Jm^{-3}, 16.9 m

(17) Distinguish between a free and forced vortex.

(18) Show that the surface of a liquid stirred within a cylindrical vessel forming a forced vortex is a paraboloid.

(19) Determine the difference in pressure between radii of 12 cm and 6 cm of a forced vortex rotated at 1000 rpm.

Answer: 59.2 kNm^{-2}

(20) In a free cylindrical vortex of water, the pressure is found to be 200 kNm^{-2} at a radius of 6 cm and tangential velocity of 6 ms^{-1}. Determine the pressure at a radius of 12 cm.

Answer: 213.5 kNm^{-2}

(21) A cylindrical tank of radius 1 m contains a liquid of density 1100 kgm^{-3} to a depth of 1 m. The liquid is stirred by a long paddle of diameter 60 cm, the axis of which lies along the axis of the tank. Determine the kinetic energy of the liquid per unit depth when the speed of rotation of the paddle is 45 rpm.

Answer: 904 Jm^{-1}

(22) Comment on the consequence of assuming inviscid fluid flow in terms of flow through pipes and vortex motion in tanks.

Laminar flow and lubrication

3

Introduction

Flowing fluids may exhibit one of two types of flow behaviour that can be readily distinguished. In streamline or laminar flow, fluid particles move along smooth parallel paths or layers (laminae) in the direction of flow with only minor movement across the streamlines caused by diffusion.

Over the centuries, the existence of laminar and turbulent flow has been studied extensively by many prominent scientists. In 1839, it was first noted by the German hydraulics engineer Gotthilf Heinrich Ludwig Hagen that laminar flow ceased when the velocity of a flowing fluid increased beyond a certain limit. With much work on the subject over the following three decades, Hagen finally concluded in 1869 that the transition from laminar to turbulent flow was dependent on velocity, viscosity and pipe diameter. Around the same time, French physician and physicist, Jean Louis Marie Poiseuille, whilst researching the effects of blood flow in veins, reported similar independent work on the viscosity and pressure drop of water in capillaries and reached the same mathematical conclusions as Hagen.

It was not until 1883 that British scientist Osborne Reynolds showed by both dimensional analysis and experiment that the transition depends on a single dimensionless parameter which bears his name. This is given by

$$Re = \frac{\rho v d}{\mu}$$

where Re is the Reynolds number, ρ is the density, v the average velocity and μ the viscosity of the fluid, and d is the diameter of pipe.

Reynolds' experiments involved injecting a trace of coloured liquid into the flow of water in a horizontal glass tube. At low flowrates the coloured liquid was observed to remain as discrete filaments along the tube axis, indicating flow in parallel streams. At increased flow, oscillations were observed in the filaments which eventually broke up and dispersed across the tube. The critical value of Reynolds number for the break-up of laminar flow was found to be about 2000 while turbulent flow was not found to occur until above 4000. The so-called transition zone, which lies between Reynolds numbers of 2000 and 4000, was found to be a region of fluid streamline instability and unpredictability.

The majority of mathematical problems involving laminar flow tend to be straightforward. The basis of most calculations involves applying a simple equilibrium force balance to the flowing fluid with fully developed laminar flow. Without considering inertial forces, the acceleration of particles in the fluid caused by the pressure gradient is retarded by the viscous shear stresses set up by the velocity gradient perpendicular to the direction of flow. This approach forms the basis of the independent work of both Hagen and Poiseuille, from whence it is possible to determine important flow details such as velocity profiles and rates of flow through pipes, gaps and channels as well as the viscous and lubricating effects of fluids in bearings.

3.1 Reynolds number equations

Establish equations for Reynolds number for the flow of a fluid with average velocity v in a pipe of inside diameter d, in terms of kinematic viscosity, volumetric flowrate, mass flowrate and mass loading.

Solution

The Reynolds number is an important dimensionless parameter and is valuable for identifying whether the flow of a fluid is either laminar or turbulent. The Reynolds number is traditionally given as

$$Re = \frac{\rho v d}{\mu}$$

It can be alternatively expressed in terms of kinematic viscosity which is

$$\nu = \frac{\mu}{\rho}$$

to give a Reynolds number expression of

$$Re = \frac{vd}{\nu}$$

Since the average velocity v is related to volumetric flowrate by

$$v = \frac{4Q}{\pi d^2}$$

then the Reynolds number can be given by

$$Re = \frac{4\rho Q}{\mu \pi d}$$

Likewise, since volumetric flowrate is related to mass flowrate by density

$$Q = \frac{m}{\rho}$$

then Reynolds number is

$$Re = \frac{4m}{\mu \pi d}$$

Finally, since mass loading is related to mass flowrate by flow area

$$L = \frac{4m}{\pi d^2}$$

The Reynolds number is therefore

$$Re = \frac{Ld}{\mu}$$

It should be noted that the Reynolds number is of primary importance in considering the flow of fluids through pipes and open channels, and around objects. As a dimensionless parameter the Reynolds number ($Re = \rho v d/\mu$) represents the ratio of inertial to viscous forces in a flowing fluid and is directly associated with the boundary surface over which the fluid passes. For laminar flow in circular pipes ($Re < 2000$), the viscous forces dominate while the inertial forces are of little significance. In turbulent flow the reverse is true with the laminar sublayer being destroyed and the viscous forces being of little significance.

The Reynolds number is also of importance in considering the viscous drag on submerged objects. For a Reynolds number below a value of 0.5 the precise shape of the object is of less importance than the viscosity of the fluid and the velocity of flow. For such conditions it is possible to derive a relationship for the drag on objects in which the inertial effects are not included. For Reynolds numbers above 0.5, however, the reverse is again true — the influence of viscosity diminishes while the influence of inertia progressively dominates with increasing values of *Re*.

3.2 Laminar boundary layer

Determine the furthest distance into a pipe with an inside diameter of 8 mm and a well-rounded entrance at which fully developed laminar flow begins.

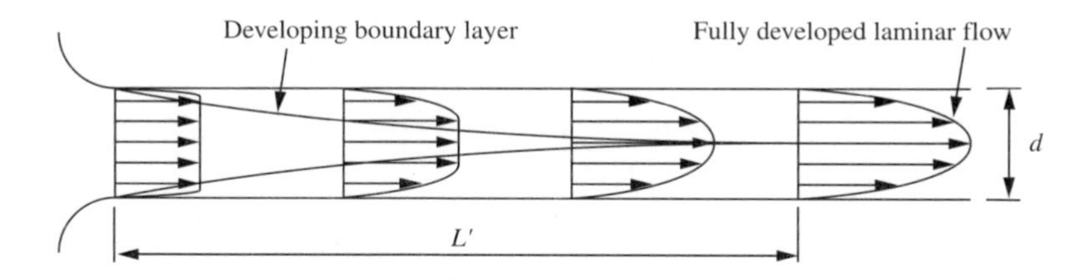

Solution

When a fluid with uniform velocity flow enters a well-rounded pipe, entrance particles close to the pipe wall are brought to rest due to the viscous properties of the fluid. This is known as the no-slip condition. A steep velocity gradient then exists in the fluid varying from zero at the pipe wall to the uniform velocity in the bulk of the fluid. As the fluid moves further into the pipe, the viscous retardation of the particles in adjacent layers gradually increases in thickness. For the steady flow through the pipe, the fluid near the centre of the pipe accelerates until an equilibrium condition is reached for fully developed laminar flow with a parabolic variation of velocity with pipe radius. The region where the velocity is changing and one where it is uniform is known as the laminar boundary layer. Since the uniform velocity is approached asymptotically, the edge of the boundary layer is defined as a point where the fluid velocity reaches 99% of its theoretical maximum value. The region prior to the section at which laminar flow is fully developed is known as the entrance transition length L', for which a theoretical formula proposed by Henry L. Langhaar in 1942 and supported by good experimental agreement is given by

$$L' = 0.058 Re\, d$$

Since the maximum entrance transition length occurs for a Reynolds number of 2000 then

$$L' = 0.058 \times 2000 \times 0.008$$

$$= 0.928 \text{ m}$$

The furthest distance the fluid can flow into the 8 mm inside diameter pipe before fully developed laminar flow can exist is 0.928 m.

3.3 Velocity profile in a pipe

Derive an expression for the local velocity of a fluid flowing with fully developed laminar flow through a horizontal pipe of radius R and sketch the velocity profile.

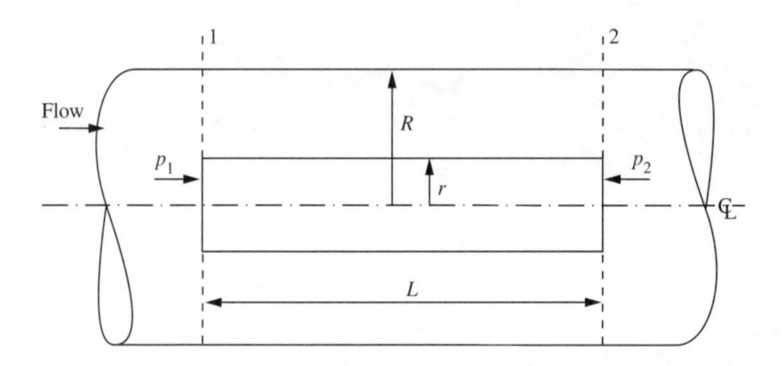

Solution

Consider a cylindrical element of the fluid to be well away from the wall on the centre line of the pipe where the difference of pressure force across the ends provides the driving force to overcome the friction on its outer surface. For uniform flow, an equilibrium force balance on the element is therefore

$$(p_1 - p_2)\pi r^2 = \tau 2\pi r \, \mathrm{d}L$$

That is

$$p_1 \pi r^2 - \left(p_1 + \frac{\delta p}{\delta L} \mathrm{d}L \right) \pi r^2 = \tau 2\pi r \, \mathrm{d}L$$

The viscous shear stress is given by

$$\tau = \mu \frac{\mathrm{d}v_x}{\mathrm{d}r}$$

Let

$$\frac{\delta p}{\delta L} = \frac{\Delta p}{L}$$

and applying the no-slip condition at the wall ($v_x = 0$ at $r = R$) then

$$\int_0^{v_x} \mathrm{d}v_x = \frac{-1}{2\mu}\frac{\Delta p}{L}\int_R^r r\,\mathrm{d}r$$

where Δp is the pressure drop over the length of element ($p_1 - p_2$). Integration gives

$$v_x = \frac{-1}{2\mu}\frac{\Delta p}{L}\left[\frac{r^2}{2}\right]_R^r$$

$$= \frac{1}{4\mu}\frac{\Delta p}{L}(R^2 - r^2)$$

The velocity of the fluid therefore has a parabolic variation with pipe radius.

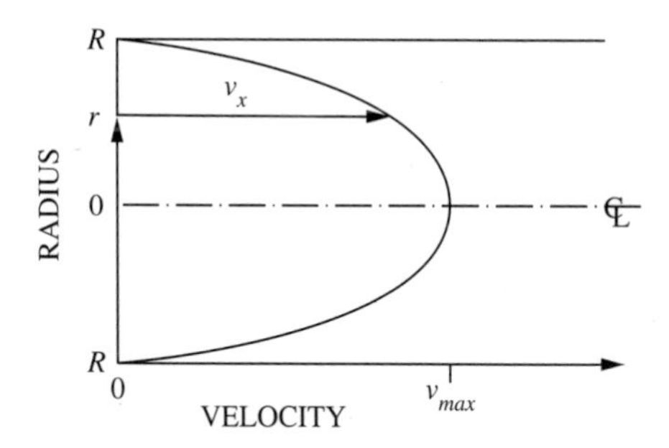

Note that the maximum velocity of the fluid, v_{max}, occurs at the furthest distance from the wall ($r = 0$). That is

$$v_{max} = \frac{1}{4\mu}\frac{\Delta p}{L}R^2$$

3.4 Hagen-Poiseuille equation for laminar flow in a pipe

A process vessel is to be supplied with glycerol of SG 1.26 and viscosity 1.2 Nsm^{-2} from a storage tank 60 m away. A pipe with an inside diameter of 12.6 cm is available together with a pump capable of developing a delivery pressure of 90 kNm^{-2} over a wide range of flows. Determine the glycerol delivery rate using this equipment. Confirm that the flow is laminar.

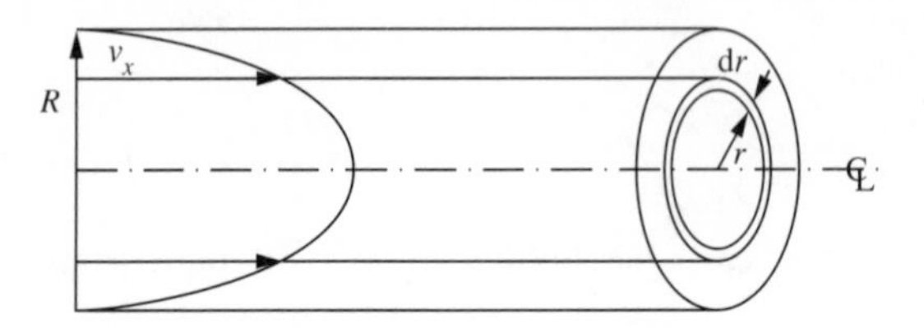

Solution

To determine the rate of flow of the fluid through the pipe with laminar flow consider a thin annular element between radius r and $r+\mathrm{d}r$ for which the velocity can be taken as constant. The elemental volumetric flowrate is therefore

$$\mathrm{d}Q = v_x 2\pi r\,\mathrm{d}r$$

The total rate of flow can therefore be found by integrating across the entire pipe radius

$$\int_0^Q \mathrm{d}Q = \int_0^R v_x\, 2\pi r\,\mathrm{d}r$$

Since the local velocity is related to radius by

$$v_x = \frac{1}{4\mu}\frac{\Delta p}{L}(R^2 - r^2)$$ (see Problem 3.3, page 66)

the integration is therefore

$$\int_0^Q \mathrm{d}Q = \frac{\pi}{2\mu}\frac{\Delta p}{L}\int_0^R (R^2 - r^2) r\,\mathrm{d}r$$

which gives

$$Q = \frac{\pi}{2\mu}\frac{\Delta p}{L}\left[\frac{R^2 r^2}{2} - \frac{r^4}{4}\right]_0^R$$

$$= \frac{\pi}{8\mu}\frac{\Delta p}{L}R^4$$

This equation was derived independently by both Hagen in 1839 and Poiseuille in 1840 and is known as the Hagen-Poiseuille equation. The rate of flow is therefore

$$Q = \frac{\pi}{8 \times 1.2} \times \frac{90 \times 10^3}{60} \times \left(\frac{0.126}{2}\right)^4$$

$$= 0.00773 \text{ m}^3\text{s}^{-1}$$

As a check for laminar flow, the Reynolds number is

$$Re = \frac{4\rho Q}{\mu \pi d}$$

$$= \frac{4 \times 1260 \times 0.00773}{1.2 \times \pi \times 0.126}$$ (see Problem 3.1, page 63)

$$= 82$$

and is below 2000, therefore confirming laminar flow.

3.5 Pipe diameter for laminar flow

In the design of a small-scale bioprocess plant, a shear sensitive Newtonian liquid of density 1100 kgm^{-3} and viscosity 0.015 Nsm^{-2} is to be transported along a length of pipe at a rate of 4 litres per minute. Determine the diameter of the pipe required if the pressure drop along the pipe is not to exceed 100 Nm^{-2} per metre length. Comment on the value of the Reynolds number.

Solution

Assuming laminar flow, the Hagen-Poiseuille equation is

$$Q = \frac{\pi}{8\mu}\frac{\Delta p}{L}R^4$$

(see Problem 3.4, page 68)

Rearranging, the pipe radius is

$$R = \left(\frac{8\mu Q}{\pi \dfrac{\Delta p}{L}}\right)^{\frac{1}{4}}$$

$$= \left(\frac{8 \times 0.015 \times \dfrac{0.004}{60}}{\pi \times 100}\right)^{\frac{1}{4}}$$

$$= 0.0126 \text{ m}$$

The diameter of the pipe is therefore 25.2 mm. To confirm laminar flow, the Reynolds number is

$$Re = \frac{4\rho Q}{\pi \mu d}$$

$$= \frac{4 \times 1100 \times \dfrac{0.004}{60}}{\pi \times 0.015 \times 0.0252}$$

(see Problem 3.1, page 63)

$$= 247$$

The flow is therefore laminar with a Reynolds number of 247.

3.6 Laminar flow through a tapered tube

A lubricating oil of viscosity 0.03 Nsm^{-2} is delivered to a machine at a rate of 10^{-7} m^{3}s^{-1} through a convergent, tapered tube of length 50 cm with an upstream diameter of 10 mm and downstream diameter of 5 mm. Determine the pressure differential which will maintain the flow. Entrance and exit losses, and inertia effects due to the change in velocity in the tube, may be neglected.

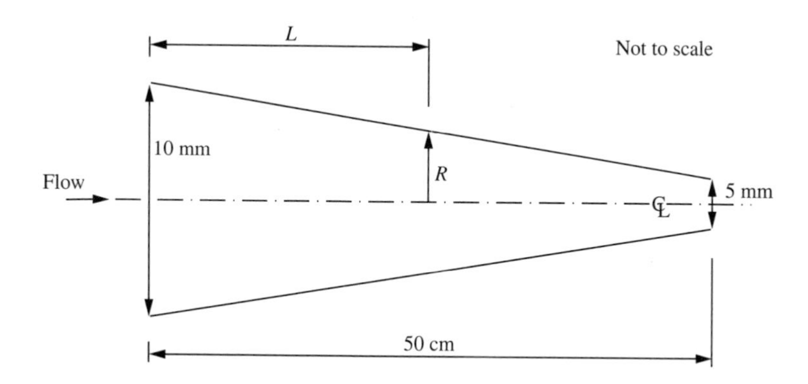

Solution

For laminar flow, the rate of flow is given by the Hagen-Poiseuille equation

$$Q = \frac{\pi}{8\mu}\frac{\Delta p}{L}R^4$$ (see Problem 3.4, page 68)

Over the length of the tapered tube, the differential pressure Δp increases with decreasing tube radius. From geometry

$$R = 0.005(1 - L)$$

where R and L are both measured in metres. Substituting for R, the total pressure differential is therefore obtained by integrating over the length of tube. That is

$$\Delta p = \frac{8\mu Q}{\pi 0.005^4}\int_0^{0.5}\frac{\mathrm{d}L}{(1 - L)^4}$$

This is made easier using the substitution

$$u = 1 - L$$

Thus

$$du = -dL$$

The integration then becomes

$$\Delta p = \frac{-8\mu Q}{\pi 0.005^4} \int \frac{du}{u^4}$$

Integrating with respect to u gives

$$\Delta p = \frac{8\mu Q}{\pi 0.005^4} \frac{u^{-3}}{3}$$

That is

$$\Delta p = \frac{8\mu Q}{3\pi 0.005^4} \left[\frac{1}{(1-L)^3} \right]_0^{0.5}$$

$$= \frac{8 \times 0.03 \times 10^{-7}}{3 \times \pi \times 0.005^4} \times \left(\frac{1}{(1-0.5)^3} - 1 \right)$$

$$= 28.5 \text{ Nm}^{-2}$$

The pressure differential is found to be 28.5 Nm^{-2}.

3.7 Relationship between average and maximum velocity in a pipe

Deduce the relationship between the average and maximum velocity for a fluid flowing with fully developed laminar flow through a horizontal pipe.

Solution

The average velocity of the fluid Q can be determined from the total rate of flow through the total cross-sectional area a available for flow

$$v = \frac{Q}{a}$$

Since the flow Q is given by the Hagen-Poiseuille equation (see Problem 3.4, page 68), then

$$v = \frac{\frac{\pi}{8\mu}\frac{\Delta p}{L}R^4}{\pi R^2}$$

$$= \frac{1}{8\mu}\frac{\Delta p}{L}R^2$$

Also, since the maximum velocity occurs at the furthest point from the wall ($r = 0$) (see Problem 3.3, page 66)

$$v_{max} = \frac{1}{4\mu}\frac{\Delta p}{L}R^2$$

then the ratio of average velocity to maximum velocity is

$$\frac{v}{v_{max}} = \frac{\frac{1}{8\mu}\frac{\Delta p}{L}R^2}{\frac{1}{4\mu}\frac{\Delta p}{L}R^2}$$

$$= \frac{1}{2}$$

That is, the average velocity of a flowing fluid with laminar flow is half the maximum velocity.

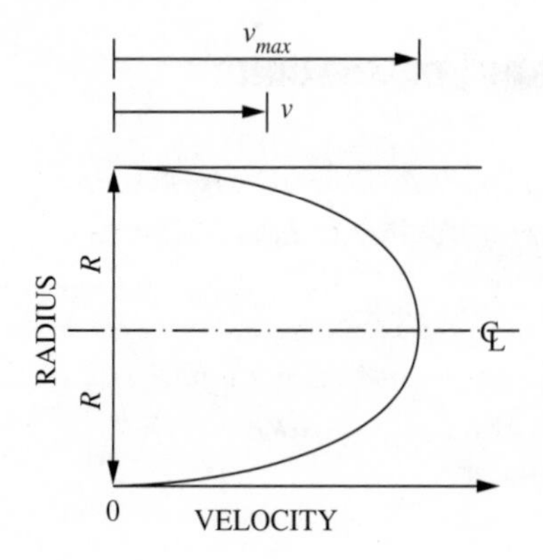

Note that the average velocity can be alternatively expressed in terms of pipe diameter as

$$v = \frac{1}{32\mu}\frac{\Delta p}{L}d^2$$

and maximum velocity as

$$v_{max} = \frac{1}{16\mu}\frac{\Delta p}{L}d^2$$

3.8 Relationship between local and maximum velocity in a pipe

Fuel oil of density 900 kgm^{-3} and viscosity 0.1 Nsm^{-2} flows through a pipe of inside diameter 100 mm. The velocity at the centre line is 2 ms^{-1}. Derive an expression relating the local to maximum velocity of a fluid flowing with fully developed laminar flow through a horizontal pipe, and determine the shear stress in the fuel oil at the pipe wall and the pressure drop per unit length of pipe.

Solution

The local velocity is given by

$$v_x = \frac{1}{4\mu}\frac{\Delta p}{L}(R^2 - r^2)$$ (see Problem 3.3, page 66)

and the maximum velocity occurs at the furthest point from the pipe wall ($r = 0$).

$$v_{max} = \frac{1}{4\mu}\frac{\Delta p}{L}R^2$$

Therefore

$$\frac{v_x}{v_{max}} = \frac{\frac{1}{4\mu}\frac{\Delta p}{L}(R^2 - r^2)}{\frac{1}{4\mu}\frac{\Delta p}{L}R^2}$$

$$= 1 - \left(\frac{r}{R}\right)^2$$

Since the viscous shear stress is

$$\tau = \mu\frac{dv_x}{dr}$$

the velocity gradient is therefore obtained by differentiation to give

$$\frac{dv_x}{dr} = -2v_{max}\frac{r}{R^2}$$

Thus

$$\tau = -2\mu v_{max}\frac{r}{R^2}$$

At the pipe wall ($r = R$), the shear stress is therefore

$$\tau_w = \frac{-2\mu v_{max}}{R}$$

$$= \frac{-2 \times 0.1 \times 2}{0.05}$$

$$= (-)8 \text{ Nm}^{-2}$$

The negative sign indicates that the shear stress is in the opposite direction to flow. To verify laminar flow, the Reynolds number is

$$Re = \frac{\rho v d}{\mu}$$

$$= \frac{\rho \dfrac{v_{max}}{2} d}{\mu}$$

$$= \frac{900 \times \dfrac{2}{2} \times 0.1}{0.1}$$

$$= 900$$

which is less than 2000. Finally, the pressure drop along the pipe is obtained from

$$\frac{\Delta p}{L} = \frac{4\mu v_{max}}{R^2}$$

$$= \frac{4 \times 0.1 \times 2}{0.05^2}$$

$$= 320 \text{ Nm}^{-2}\text{m}^{-1}$$

The pipe wall shear stress is found to be 8 Nm^{-2} and pressure drop to be 320 Nm^{-2} per metre length of pipe.

3.9 Maximum pipe diameter for laminar flow

A light oil of viscosity 0.032 Nsm^{-2} and SG 0.854 is to be transferred from one vessel to another through a horizontal pipe. If the pressure drop along the pipe is not to exceed 150 Nm^{-2} per metre length to ensure a satisfactory flow and flow is always to be laminar, determine the maximum possible internal diameter of the pipe.

Solution

The maximum value of Reynolds number at which laminar flow can exist is 2000. That is

$$Re = \frac{\rho v d}{\mu}$$

$$= 2000$$

where the average velocity is given by

$$v = \frac{1}{8\mu}\frac{\Delta p}{L}R^2$$

(see Problem 3.7, page 73)

Substituting, the Reynolds number is therefore

$$Re = \frac{\rho\left(\frac{1}{8\mu}\frac{\Delta p}{L}\left(\frac{d}{2}\right)^2\right)d}{\mu}$$

$$= \frac{\rho}{32\mu^2}\frac{\Delta p}{L}d^3$$

or rearranging in terms of pipe diameter

$$d = \left(\frac{32\mu^2 Re}{\rho\frac{\Delta p}{L}}\right)^{\frac{1}{3}}$$

$$= \left(\frac{32 \times 0.032^2 \times 2000}{854 \times 150}\right)^{\frac{1}{3}}$$

$$= 0.08 \text{ m}$$

The maximum inside diameter is found to be 8 cm.

3.10 Vertical pipe flow

A viscous liquid, of density 1000 kgm^{-3} and viscosity 0.1 Nsm^{-2}, flows between two storage tanks through 10 m of vertical pipe which has an inside diameter of 5 cm. The rate of flow is controlled by a valve at a rate of 14 m^{3}h^{-1}. Determine the pressure drop across the valve when the difference in levels between the two storage tanks is 12 m. Both tanks are open to atmosphere. Neglect losses in the pipe fittings.

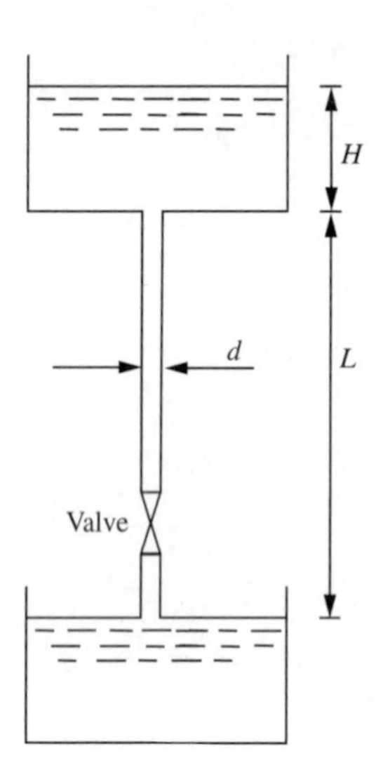

Solution

Gravity has no effect for laminar flow in a horizontal pipe. For vertical flow, however, gravitational effects are in the form of hydrostatic pressure where, in this case, the pressure drop across the valve is

$$\Delta p_{valve} = \rho g (L + H) - \frac{128 Q \mu L}{\pi d^4}$$

$$= 1000 \times g \times (10 + 2) - \frac{128 \times \frac{14}{3600} \times 0.1 \times 10}{\pi \times 0.05^4}$$

$$= 92{,}368 \text{ Nm}^{-2}$$

The pressure drop across the valve is found to be 92.4 kNm^{-2}.

3.11 Film thickness in a channel

Derive expressions for the velocity distribution, total flow and film thickness of a Newtonian liquid flowing under gravity down a surface inclined at an angle θ to the horizontal. Assume that the flow is laminar and fully developed.

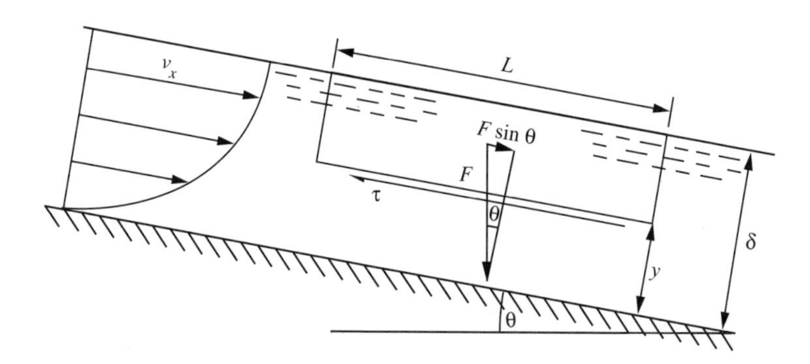

Solution

Assuming steady flow, a negligible shear stress at the gas-liquid interface and a uniform pressure in the direction of flow (x-direction), an equilibrium force balance on the element gives

$$WL(\delta - y)\rho g \sin\theta = \tau WL$$

where the shear stress is given by

$$\tau = \mu \frac{dv_x}{dy}$$

The velocity profile through the film can be found using the no-slip boundary condition (y =0 at v_x =0). Therefore

$$\int_0^{v_x} dv_x = \frac{\rho g \sin\theta}{\mu}\int_0^{y}(\delta - y)dy$$

Integrating

$$v_x = \frac{\rho g \sin\theta}{\mu}\left(\delta y - \frac{y^2}{2}\right)$$

The volume flowrate of the liquid is given by

$$\int_0^Q \mathrm{d}Q = \int_0^\delta W v_x \,\mathrm{d}y$$

Therefore substituting for v_x

$$Q = \frac{W\rho g \sin\theta}{\mu} \int_0^\delta \left(\delta y - \frac{y^2}{2} \right) \mathrm{d}y$$

Integrating

$$Q = \frac{W\rho g \sin\theta}{\mu} \left[\frac{\delta y^2}{2} - \frac{y^3}{6} \right]_0^\delta$$

$$= \frac{W\rho g \sin\theta}{3\mu} \delta^3$$

Rearranging, the film thickness is therefore related to flowrate by

$$\delta = \left(\frac{3\mu Q}{W\rho g \sin\theta} \right)^{\frac{1}{3}}$$

The film thickness can also be related to the average velocity by

$$v = \frac{Q}{\delta W}$$

$$= \frac{W\rho g \sin\theta}{3\mu\delta W} \delta^3$$

$$= \frac{\rho g \sin\theta}{3\mu} \delta^2$$

Rearranging, the thickness of the film of liquid is therefore

$$\delta = \left(\frac{3\mu v}{\rho g \sin\theta} \right)^{\frac{1}{2}}$$

This equation has been found by experiment to hold for a Reynolds number ($Re = \rho v\delta/\mu$) less than 500.

3.12 Flow down an inclined plate

In the first of two experiments water, of density 1000 kgm^{-3} and viscosity 0.001 Nsm^{-2}, was pumped to the top of a flat plate and evenly distributed along one side of it. The plate was angled at approximately 60° to the horizontal and the flowrate was 3×10^{-6} m^{3}s^{-1} per metre length of plate. It was found that the film thickness was 1.016×10^{-4} m. In the second experiment, a different liquid of density 1500 kgm^{-3} was pumped to the top of the plate at 3×10^{-6} m^{3}s^{-1} per metre length of plate and the film thickness was found to be 1.25×10^{-4} m. Determine the viscosity of the second liquid and the exact angle of the plate.

Solution

The rate of flow of liquid down the plate is given by

$$Q = \frac{W\rho g \sin\theta}{3\mu}\delta^3$$

(see Problem 3.11, page 79)

For the first experiment involving water, the exact angle of inclination is found by rearranging the equation to

$$\sin\theta = \frac{3Q\mu}{W\rho g\delta^3}$$

$$= \frac{3\times3\times10^{-6}\times1\times10^{-3}}{1\times1000\times g\times(1.016\times10^{-4})^3}$$

$$= 0.875$$

The exact angle of inclination was therefore

$$\theta = 61°$$

For the second experiment, the viscosity is found by rearranging the equation for flow to

$$\mu = \frac{W\rho g \sin\theta}{3Q}\delta^3$$

$$= \frac{1\times1000\times g\times0.875}{3\times3\times10^{-6}}\times(1.25\times10^{-4})^3$$

$$= 2.795\times10^{-3}\ \text{Nsm}^{-2}$$

The exact angle was 61° and the viscosity of the second liquid was found to be 2.795×10^{-3} Nsm^{-2}.

3.13 Flow down a vertical wire

A film of a Newtonian fluid flows down the outside of a stationary vertical wire which has a radius R. Deduce an expression for the local velocity of the fluid, v_z*, down the wire.*

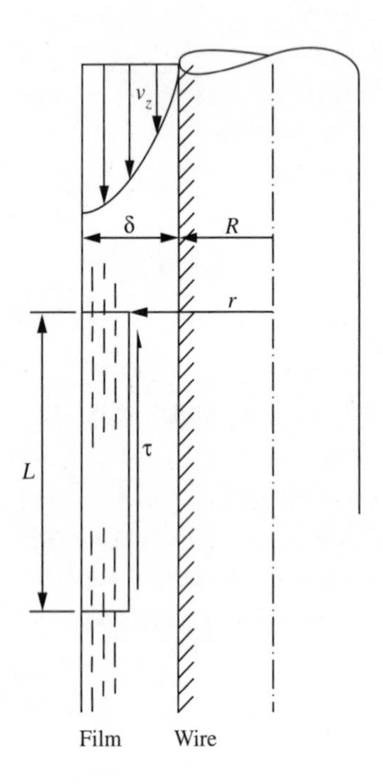

Solution

Consider a force balance on an element of the fluid between radii r and $R + \delta$ assuming no shear effects at the free surface. Then

$$\rho g L\pi((R+\delta)^2 - r^2) = \tau 2\pi r L$$

where the viscous shear stress is given by

$$\tau = \mu \frac{\mathrm{d}v_z}{\mathrm{d}r}$$

Substituting and noting that the fluid is stationary (no-slip condition) at the wire surface ($v_z = 0$ at $r = R$), the local velocity at radius r measured from the axis of the wire can be found from

$$\int_0^{v_z} \mathrm{d}v_z = \frac{\rho g}{2\mu}\int_R^r \left(\frac{(R+\delta)^2}{r} - r\right)\mathrm{d}r$$

$$= \frac{\rho g}{2\mu}\left((R+\delta)^2 \int_R^r \frac{\mathrm{d}r}{r} - \int_R^r r\mathrm{d}r\right)$$

Integrating with respect to r gives

$$v_z = \frac{\rho g}{2\mu}\left((R+\delta)^2 \log_e [r]_R^r - \left[\frac{r^2}{2}\right]_R^r\right)$$

$$= \frac{\rho g}{4\mu}\left(2(R+\delta)^2 \log_e\left(\frac{r}{R}\right) + (R^2 - r^2)\right)$$

While the velocity of the fluid is zero at the wire surface, it is at its greatest at the free surface. The average velocity, rate of flow and shear stress within the fluid can similarly be determined and are useful results when considering the process of wire coating. This involves the drawing of a wire vertically up through a bath of viscous liquid taking with it a film of the liquid. In practice, however, the coating liquid — such as a polymer melt — is not usually likely to exhibit Newtonian behaviour and, on drying, density change and shrinkage must also be taken into consideration.

3.14 Flow and local velocity through a gap

Derive an expression for the local velocity and total flow of a Newtonian fluid with fully developed laminar flow between two flat horizontal plates separated by a distance 2H.

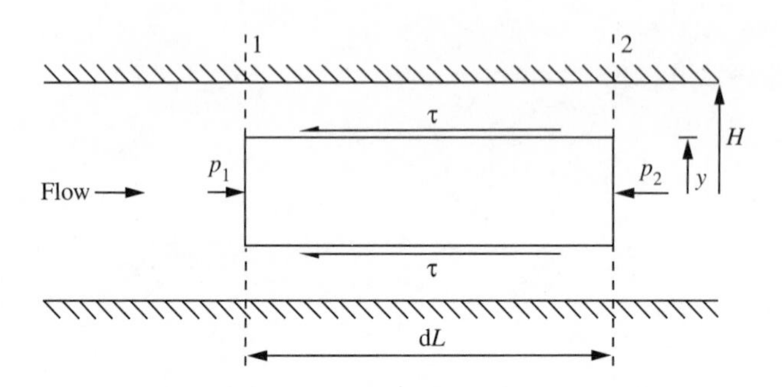

Solution

Assuming that the plates are sufficiently wide that edge effects can be neglected (that is, $W \gg 2H$), an equilibrium force balance on the element is

$$2yW(p_1 - p_2) = 2\tau W\mathrm{d}L$$

That is

$$2yW\left(p_1 - \left(p_1 + \frac{\delta p}{\delta L}\mathrm{d}L\right)\right) = 2\tau W\mathrm{d}L$$

where the shear stress is given by

$$\tau = \mu \frac{\mathrm{d}v_x}{\mathrm{d}y}$$

Let

$$\frac{\delta p}{\delta L} = \frac{\Delta p}{L}$$

The local velocity v_x at a distance y can therefore be found applying the no-slip boundary condition ($v_x = 0$ at $y = H$). That is

$$\int_0^{v_x} \mathrm{d}v_x = \frac{-1}{\mu}\frac{\Delta p}{L}\int_H^y y\mathrm{d}y$$

Integration with respect to y gives

$$v_x = \frac{-1}{\mu}\frac{\Delta p}{L}\left[\frac{y^2}{2}\right]_H^y$$

$$= \frac{1}{2\mu}\frac{\Delta p}{L}(H^2 - y^2)$$

The velocity of the fluid therefore has a parabolic variation with distance between the plates, with the maximum velocity at $y = 0$.

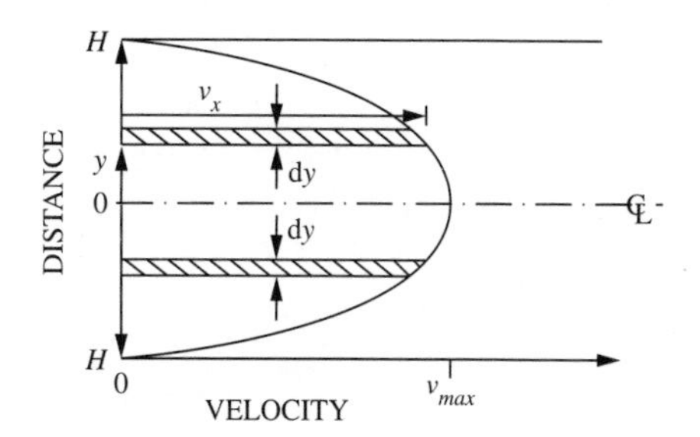

The total flow of fluid can be determined by considering the sum of the elemental flows through a narrow slit of thickness dy. From symmetry about the mid-point between the plates there are two such narrow slits. Thus

$$\int_0^Q \mathrm{d}Q = 2W\int_0^H v_x \mathrm{d}y$$

Substituting for v_x

$$\int_0^Q \mathrm{d}Q = \frac{W}{\mu}\frac{\Delta p}{L}\int_0^H (H^2 - y^2)\mathrm{d}y$$

and integration with respect to y gives

$$Q = \frac{W}{\mu}\frac{\Delta p}{L}\left[yH^2 - \frac{y^3}{3}\right]_0^H$$

$$= \frac{2W}{3\mu}\frac{\Delta p}{L}H^3$$

In addition to assuming that edge effects can be neglected, for incompressible viscous flow of Newtonian fluids through gaps or between plates it is also assumed that there is no pressure gradient across the fluid layer and that gravitational forces are negligible.

3.15 Relationship between local and average velocity through a gap

Derive an expression relating the local velocity of a Newtonian fluid flowing with fully developed laminar flow between two horizontal parallel plates separated by a distance 2H in terms of the average velocity.

Solution

The total flow is related to average velocity by

$$Q = 2WHv$$

The average velocity is therefore obtained by substituting an expression for laminar flow between the plates (see Problem 3.14, page 84)

$$Q = \frac{2W}{3\mu}\frac{\Delta p}{L}H^3$$

to give

$$v = \frac{\frac{2W}{3\mu}\frac{\Delta p}{L}H^3}{2WH}$$

$$= \frac{1}{3\mu}\frac{\Delta p}{L}H^2$$

The local velocity v_x between the plates is given by

$$v_x = \frac{1}{2\mu}\frac{\Delta p}{L}(H^2 - y^2)$$

(see Problem 3.14, page 84)

Therefore

$$\frac{v_x}{v} = \frac{\frac{1}{2\mu}\frac{\Delta p}{L}(H^2 - y^2)}{\frac{1}{3\mu}\frac{\Delta p}{L}H^2}$$

The local velocity is therefore related to average velocity by

$$v_x = \frac{3v}{2}\left(1 - \left(\frac{y}{H}\right)^2\right)$$

where y is the distance measured away from the middle of the gap.

3.16 Relationship between average and maximum velocity through a gap

Deduce the ratio of maximum velocity to average velocity for a Newtonian fluid flowing with laminar flow through a gap between two flat horizontal plates.

Solution

The maximum velocity occurs at the furthest point from either surface ($y = 0$) (see Problem 3.14, page 84). That is

$$v_{max} = \frac{1}{2\mu}\frac{\Delta p}{L}H^2$$

Since the average velocity is given by

$$v = \frac{1}{3\mu}\frac{\Delta p}{L}H^2 \qquad \text{(see Problem 3.15, page 86)}$$

Then

$$\frac{v_{max}}{v} = \frac{\frac{1}{2\mu}\frac{\Delta p}{L}H^2}{\frac{1}{3\mu}\frac{\Delta p}{L}H^2}$$

$$= \frac{3}{2}$$

The maximum velocity is one and a half times the average velocity through the gap.

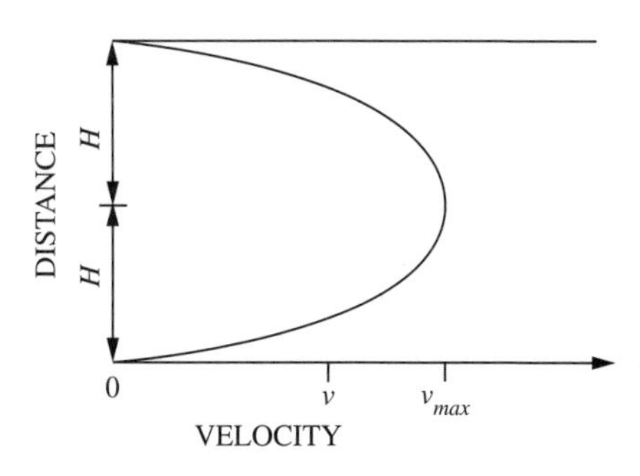

3.17 Shear stress for flow through a gap

Oil, with a viscosity of 1.5 Nsm^{-2}, flows with fully developed laminar flow through a gap formed by two horizontal parallel plates set 20 mm apart. Determine the magnitude and direction of the shear stresses that act on the plates when the average velocity is 0.5 ms^{-1}.

Solution

The velocity of the fluid is given by

$$v_x = \frac{3v}{2}\left(1 - \left(\frac{y}{H}\right)^2\right)$$

(see Problem 3.15, page 86)

The velocity gradient is therefore

$$\frac{dv_x}{dy} = \frac{d}{dy}\left(\frac{3v}{2}\left(1 - \left(\frac{y}{H}\right)^2\right)\right)$$

$$= \frac{-3yv}{H^2}$$

The viscous shear stress at either plate surface (wall) where $y = H$ is therefore

$$\tau = \tau_w$$

$$= \mu\left(\frac{dv_x}{dy}\right)_w$$

$$= -\mu\,\frac{3v}{H}$$

$$= -1.5 \times \frac{3 \times 0.5}{\frac{0.02}{2}}$$

$$= -225\ \text{Nm}^{-2}$$

The shear stress at the wall is 225 Nm^{-2} and acts in the opposite direction to flow.

3.18 Flat disc viscometer

A device used to measure the viscosity of a viscous liquid consists of a flat disc which rotates on a flat surface between which is sandwiched the liquid under investigation. The disc has a diameter of 5 cm and produces a shear stress of 400 Nm^{-2} for a rotational speed of 600 rpm. Determine the viscosity of the liquid and the torque on the rotating disc if the clearance between the disc and surface is 2 mm.

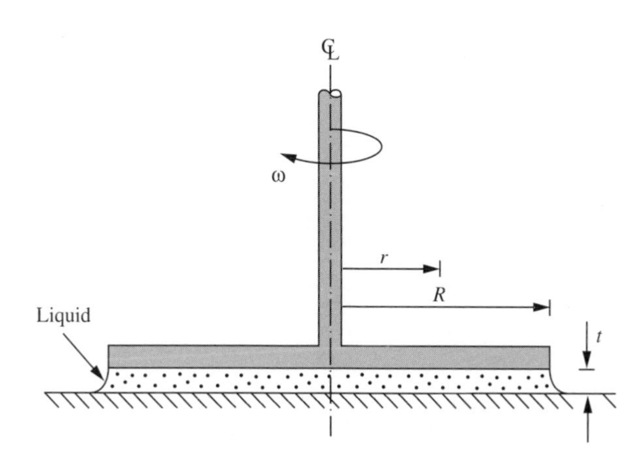

Solution

The viscosity of liquids can be determined experimentally using a device known as a viscometer. There are various types of viscometer available and all are based on the principles of laminar flow. The flat disc viscometer is perhaps one of the simplest forms, and can be used for quick estimates of viscosity. For the liquid sandwiched between the surface and the rotating disc, the shear stress is given by

$$\tau = \mu \frac{dv_x}{dy}$$

or in terms of angular velocity and clearance between disc and plate

$$\tau = \mu \frac{\omega r}{t}$$

where

$$\omega = 2\pi N$$

$$= 2 \times \pi \times \frac{600}{60}$$

$$= 62.8 \text{ rad s}^{-1}$$

Rearranging in terms of viscosity

$$\mu = \frac{\tau t}{\omega r}$$

$$= \frac{400 \times 2 \times 10^{-3}}{62.8 \times 0.025}$$

$$= 0.509 \text{ Nsm}^{-2}$$

The torque is given by

$$T = 2\pi \int_0^R \tau r^2 \mathrm{d}r$$

Substituting for shear stress

$$T = \frac{2\pi\mu\omega}{t} \int_0^R r^3 \mathrm{d}r$$

Integrating

$$T = \frac{2\pi\mu\omega}{t} \left[\frac{r^4}{4} \right]_0^R$$

$$= \frac{\pi\mu\omega}{2t} R^4$$

$$= \frac{\pi \times 0.509 \times 62.8}{2 \times 2 \times 10^{-3}} \times 0.025^4$$

$$= 9.8 \times 10^{-3} \text{ Nm}$$

The viscosity and torque are found to be 0.509 Nsm^{-2} and 9.8×10^{-3} Nm.

3.19 Torque on a lubricated shaft

A shaft 100 mm in diameter rotates at 30 rps in a bearing of length 200 mm, the surfaces being separated by a film of oil 0.02 mm thick. Determine the viscous torque on the bearing and power required to overcome the frictional resistance if the viscosity of the oil is 0.153 Nsm^{-2}.

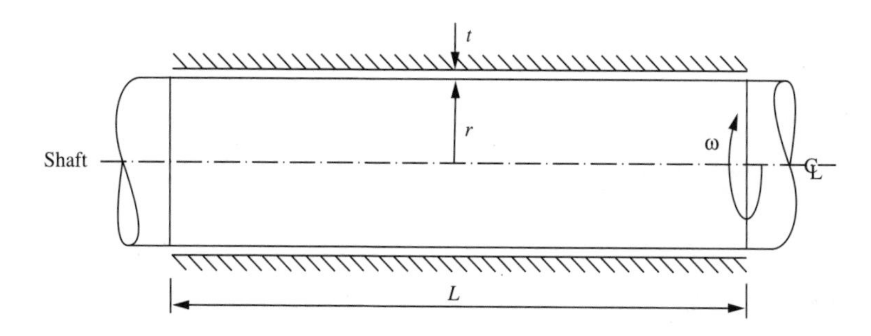

Solution

It is assumed that there are negligible end effects and that the shaft is lightly loaded so that the shaft runs concentrically. The resistance per unit area or shear stress is

$$\tau = \mu \frac{dv}{dy}$$

The oil film is very thin such that the viscous shear stress may be given in terms of angular velocity at radius r

$$\tau = \mu \frac{\omega r}{t}$$

$$= \mu \frac{2\pi N r}{t}$$

The tangential resistance (viscous drag) on the bearing is

$$F = \tau 2\pi r L$$

The resisting torque is therefore

$$T = Fr$$

$$= \tau 2\pi r L r$$

$$= \mu \frac{2\pi N r}{t} 2\pi r L r$$

$$= \mu \frac{4\pi^2 r^3 N L}{t}$$

$$= 0.153 \times \frac{4 \times \pi^2 \times 0.05^3 \times 30 \times 0.2}{2 \times 10^{-5}}$$

$$= 226 \text{ Nm}$$

The power of the shaft is

$$P = T\omega$$

$$= T 2\pi N$$

$$= 226 \times 2 \times \pi \times 30$$

$$= 42{,}600 \text{ W}$$

The torque on the bearing is found to be 226 Nm and the power required to overcome the frictional resistance is 42.6 kW.

Note that if the radial clearance between the shaft and bearing is not small, the adjustment to torque is made

$$T = \mu \frac{4\pi^2 r_1^2 r_2 N L}{r_2 - r_1}$$

where r_1 and r_2 are the radii of the shaft and bearing, respectively.

3.20 Lubricated collar bearing

The axial thrust on a shaft is taken by a collar bearing, the outside and inside diameters of which are 200 mm and 125 mm, respectively. The bearing is provided with a forced lubrication system which maintains a film of oil of thickness 0.5 mm between the bearing surfaces. At a shaft speed of 300 rpm the friction loss of power amounts to 22 W. Determine the viscosity of the oil.

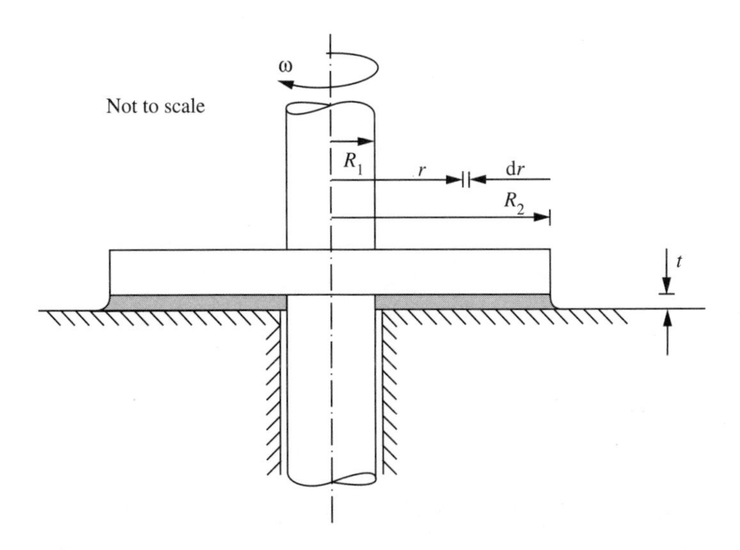

Solution

Bearings are mechanical devices used to reduce the friction between two parts in contact with one another and which move with respect to one another. Collar bearings are to be found on vertical rotating shafts and take a load on a stationary support. To reduce friction, a film of lubricating fluid separates the rotating collar and stationary support. The viscous resistance of the collar bearing can be obtained by assuming the face of the collar to be separated from the bearing surface by a thin film of oil of uniform thickness where the viscous shear stress is given by

$$\tau = \mu \frac{\mathrm{d}v}{\mathrm{d}y}$$

For the thin film of oil, the shear stress may therefore be expressed in terms of angular velocity as

$$\tau = \mu \frac{\omega r}{t}$$

The tangential viscous force changes with radius. For a thin ring at a radius r

$$= \tau 2\pi r \mathrm{d}r$$

$$= \mu \frac{\omega r}{t} 2\pi r \mathrm{d}r$$

The moment of tangential force on the ring is therefore

$$= \mu \frac{\omega r}{t} 2\pi r \mathrm{d}r \; r$$

For the whole bearing to consist of similar concentric rings, the total torque required to overcome the viscous resistance of the bearing is found by integrating

$$T = \frac{2\pi\mu\omega}{t} \int_{R_2}^{R_1} r^3 \mathrm{d}r$$

$$= \frac{\pi\mu\omega}{2t} (R_1^4 - R_2^4)$$

The power loss is related to torque by

$$P = T\omega$$

Therefore rearranging in terms of viscosity

$$\mu = \frac{2tP}{\pi\omega^2 (R_1^4 - R_2^4)}$$

$$= \frac{2 \times 0.0005 \times 22}{\pi \times \left(2 \times \pi \times \dfrac{300}{60}\right)^2 \times (0.1^4 - 0.0625^4)}$$

$$= 0.0837 \text{ Nsm}^{-2}$$

The viscosity of the oil is found to be 0.0837 Nsm^{-2}.

Further problems

(1) Explain what is meant by a Newtonian fluid.

(2) A cubic block of volume 1000 cm^3 is supported by a film of oil 0.1 mm thick on a horizontal surface. Determine the viscous drag if the block is moved across the surface with a velocity of 1 ms^{-1}. The viscosity of the oil is 0.05 Nsm^{-2}.

Answer: 5 N

(3) A block is supported by a film of oil 0.2 mm thick on a horizontal surface. The area of the oil film is 100 cm^2, the viscosity of the oil is 0.05 Nsm^{-2}. Determine the viscous drag at a velocity of 1 ms^{-1}.

Answer: 2.5 N

(4) A fluid flows at a rate Q in a pipe of diameter d_1 and has a Reynolds number Re_1. If the fluid passes to a pipe of diameter d_2, deduce the relationship between Re_1 and Re_2.

(5) Show that the local velocity of a liquid with laminar flow along a horizontal channel of width W in terms of flowrate Q and depth δ can be given by

$$v_x = \frac{3Q}{W\delta^3}\left(\delta y - \frac{y^2}{2}\right)$$

where y is the depth measured from the base of the channel.

(6) Show that the maximum velocity of a liquid with laminar flow along a horizontal channel of width W and depth δ can be given by

$$v_{max} = \frac{3Q}{2W\delta}$$

(7) Sketch the velocity profile and write down the boundary conditions for the laminar movement of a Newtonian liquid between two large flat plates which are horizontal, parallel and 1 mm apart, if the top plate moves at +0.01 ms^{-1} and the lower plate moves at –0.01 ms^{-1} and there is no net flow of fluid between the plates.

(8) Outline the assumptions on which the flow of fluid with fully developed laminar flow through a pipe is based.

(9) Glycerol, of SG 1.26 and viscosity 1.2 Nsm^{-2}, flows between two flat plates 2 cm apart. If the pressure drop along the plates is 500 Nm^{-2} per metre length, determine the maximum velocity of the glycerol.

Answer: 0.021 ms^{-1}

(10) Show that the local velocity of a Newtonian fluid with laminar flow flowing through an annular space between two tubes can be given by

$$v_x = \frac{1}{4\mu}\frac{\Delta p}{L}\left((R^2 - y^2) - (R^2 - r^2)\frac{\log_e\left[\frac{y}{R}\right]}{\log_e\left[\frac{r}{R}\right]}\right)$$

where R is the inner radius of the outer tube, r is the outer radius of the inner tube, y is a radius between R and r, μ is the dynamic viscosity of the fluid and $\Delta p / L$ is the pressure drop per unit length through the annulus.

(11) Sketch the velocity profile of a Newtonian fluid through an annular gap with laminar flow.

(12) Sketch the velocity profiles for the flow of two immiscible liquids A and B along a horizontal channel. The viscosities and thickness of layers may be assumed to be identical and where the density of liquid A is less than the density of liquid B.

(13) Show that the flowrate of the Newtonian liquid flowing down the outside of a vertical stationary wire of radius R may be given by

$$Q = \frac{\rho g \pi}{2\mu}\left(R^2(R+\delta)^2 - \frac{R^4}{4} - (R+\delta)^4\left(\frac{3}{4} - \log_e\left(\frac{R+\delta}{R}\right)\right)\right)$$

where δ is the thickness of the liquid film.

(14) Show that the local velocity v_x of Newtonian liquid flowing with fully developed laminar flow down a plate inclined at an angle θ to the horizontal may be given by the expression

$$v_x = \frac{\rho g \sin\theta}{\mu}\left(\left(\frac{3Q\mu}{\rho g \sin\theta}\right)^{\frac{1}{3}} y - \frac{y^2}{2}\right)$$

where y is the depth measured from the surface of the plate.

(15) Derive an expression for the velocity distribution for a Newtonian fluid flowing with fully developed laminar flow through a narrow gap formed by two vertical parallel plates and sketch qualitatively the velocity and shear stress profiles.

(16) A liquid flows in laminar flow along a horizontal rectangular channel which has a width W and a depth δ, where $W >> \delta$. Show that well away from the entrance, exit and side walls of the channel, the liquid velocity along the channel is given by

$$v_x = \frac{6Q}{W\delta^3}(\delta y - y^2)$$

where Q is the volumetric rate of flow and y is the distance from the bottom of the channel.

(17) A liquid flows in laminar flow at a rate Q along a horizontal tube which has an inside diameter d. Show that under steady flow conditions and far away from the entrance to or exit from the tube, the liquid velocity v_x at a radius r is given by the expression

$$v_x = \frac{32Q}{\pi d^4}\left(\frac{d^2}{4} - r^2\right)$$

where Q is the volumetric rate of flow and d is the diameter of the tube.

(18) A cylindrical bob with a diameter of 100 mm and length 150 mm revolves within a cup with a clearance of 1 mm with the annular space filled with glycerine. If the bob rotates at 120 rpm for which a torque of 2.3 Nm is recorded, determine the viscosity of the glycerine.

Answer: 0.079 $\mathrm{Nsm^{-2}}$

(19) Water flows in laminar flow along a horizontal channel which has a width of 4 m. It is observed that the depth of water in the channel is 3 cm and that small gas bubbles, which form at the floor of the channel, take 220 seconds to reach the surface after detachment. It is also observed that the bubbles reach the water surface 7.32 m, measured in the direction of flow, from the point of formation. Determine the rate of flow of water along the channel.

Answer: 0.004 $\mathrm{m^3s^{-1}}$

Dimensional analysis

4

Introduction

Dimensional analysis is a particularly useful way of obtaining relationships between variables that predict the occurrence of natural phenomena useful for scaling up models. The techniques involved, also to be found under the titles of similitude, theory of dimensions, theory of similarity and theory of models, are based on the physical relationship between the variables being required to be dimensionally perfect. By organization of the variables, it may be possible to determine a fundamental relationship between them.

Dimensional analysis is encountered in many other branches of engineering. In terms of fluid mechanics, problems involving the fundamental dimensions of mass, length and time can be applied to all forms of fluid resistance, flow through pipes, through weirs and orifices. The techniques require identification of all the requisite variables by which a phenomenon is affected; missing or wrong variables will lead to incorrect conclusions. The requirement for forming dimensionless groups is that they should be independent of one another within the set, but all possible dimensionless groups outside the set can be formed as products of powers of the groups within the complete set.

There has long been a preoccupation with dimensional analysis employing Lord Rayleigh's method of indices of 1899 and the Π theorem of Buckingham of 1915 (although first stated in 1892 by A. Vaschy). There are limitations with these theories, however, largely associated with the need to identify the variables responsible for predicting a particular phenomenon together with the interpretation of the dimensionless groups. A number of techniques have subsequently been developed that better recognize the physical features of a particular situation. Nonetheless, dimensional analysis reasoning remains a useful tool in formulating models, particularly where the laws governing a particular phenomenon are not known.

4.1 Flow through an orifice

Deduce by the method of dimensions, an expression in dimensionless terms for the rate of flow Q of a liquid of density ρ and viscosity μ in a pipe of diameter d with an orifice of diameter d_o and pressure drop Δp.

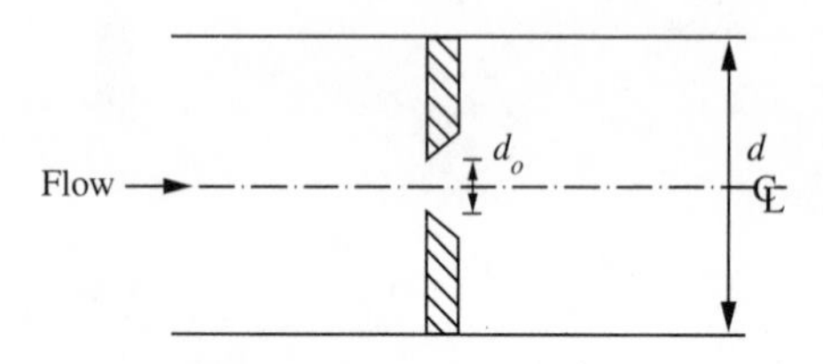

Solution

The Rayleigh method of dimensional analysis was proposed in 1899 by British physicist Lord Rayleigh (1842–1919) to determine the effect of temperature on a gas. The method, which has found wide application in engineering, involves forming an equation in which A is some function of independent variables, A_1, A_2, ... etc, which are dimensionally consistent in the form

$$A = k(A_1^a, A_2^b, ...)$$

where k is a dimensionless constant. The dimensionless groups are obtained by evaluating the exponents a, b, ... etc, and grouping those variables with the same power. For the orifice in a pipe, the relationship is therefore assumed

$$Q = k\,\Delta p^a d_o^b d^c \rho^d \mu^e$$

where the corresponding fundamental dimensions are

$$L^3T^{-1} = k(ML^{-1}T^{-2})^a L^b L^c (ML^{-3})^d (ML^{-1}T^{-1})^e$$

Equating indices for M, L and T, respectively

$$0 = a + d + e$$

$$3 = -a + b + c - 3d - e$$

$$-1 = -2a - e$$

The governing variables are pipe diameter and viscosity. A rearrangement of the indices in terms of c and e gives

$$a = \frac{1}{2} - \frac{e}{2}$$

$$b = 2 - c - e$$

$$d = -\frac{1}{2} - \frac{e}{2}$$

The power relationship is therefore

$$Q \propto \Delta p^{\frac{1}{2}} d^2 \rho^{-\frac{1}{2}} \left(\frac{\mu}{d\,\Delta p^{\frac{1}{2}} \rho^{\frac{1}{2}}} \right)^e \left(\frac{d}{d_o} \right)^c$$

The constant k therefore corresponds to a form of Reynolds number and ratio of pipe to orifice diameter. Thus

$$Q = kd^2 \sqrt{\frac{\Delta p}{\rho}}$$

This result is expected as the rate of flow depends on the velocity through the orifice. Since the pressure drop is related to head by

$$\Delta p = \rho g H$$

then the rate of flow can alternatively be expressed as

$$Q = kd^2 \sqrt{gH}$$

Note that comparing this equation with that for flow through an orifice (see Problem 5.4, page 122), it can be deduced that

$$k = \frac{\frac{C_d \pi}{4}}{\sqrt{\left(\frac{d}{d_o} \right)^4 - 1}}$$

where C_d is the discharge coefficient and a_o is the area of the orifice. Much experimental work has been done correlating C_d with Reynolds number for different ratios of pipe to orifice diameter. For Reynolds numbers above 10,000, however, C_d approaches values between 0.6 and 0.65 irrespective of d/d_o.

4.2 Flow over notches

Show by the method of dimensions that the flow of liquid Q over a notch can be given by the dimensionless groups

$$Q = kg^{\frac{1}{2}} H^{\frac{5}{2}} \left(\frac{\mu}{\rho g^{\frac{1}{2}} H^{\frac{3}{2}}} \right) \left(\frac{\sigma}{\rho g H^2} \right)$$

where H is the head, ρ is the density, μ the viscosity and σ the surface tension of the liquid, g is the gravitational acceleration and k is a constant.

Solution

For problems involving flow over notches it is known that the experimental coefficients are not constant for all heads but vary with fluid density, viscosity, and surface tension. Assuming notches to be geometrically similar — that is, they have the same characteristic dimensions where the angle θ is the same for all notches tested — the power relation is assumed to be

$$Q = kH^a g^b \mu^c \rho^d \sigma^e$$

then

$$Q \propto H^a g^b \mu^c \rho^d \sigma^e$$

The fundamental dimensions are

$$L^3 T^{-1} = (L)^a (LT^{-2})^b (ML^{-1}T^{-1})^c (ML^{-3})^d (MT^{-2})^e$$

Equating the indices for M, L and T, respectively

$$0 = c + d + e$$

$$3 = a + b - c - 3d$$

$$-1 = -2b - c - 2e$$

The governing variables are viscosity and surface tension. Obtaining the indices in terms of c and e gives

$$a = \frac{5}{2} - \frac{3}{2}c - 2e$$

$$b = \frac{1}{2} - \frac{1}{2}c - e$$

$$d = -c - e$$

Therefore

$$Q \propto g^{\frac{1}{2}} H^{\frac{5}{2}} \left(\frac{\mu}{\rho g^{\frac{1}{2}} H^{\frac{3}{2}}} \right)^{c} \left(\frac{\sigma}{\rho g H^{2}} \right)^{e}$$

The coefficient of discharge for notches therefore varies with fluid density, viscosity, head and surface tension. That is, the constant k is a function of

$$\left(\frac{\mu}{\rho g^{\frac{1}{2}} H^{\frac{3}{2}}} \right) \left(\frac{\sigma}{\rho g H^{2}} \right)$$

This is expected since the flow of fluids which have a free surface is influenced in some ways by surface tension and viscosity. This is true for open channel flow, flow over notches and weirs, and multiphase fluid flow. The precise nature of the influence is not known and generally is only appreciable for conditions of shallow flow or low head over obstructions. Surface tension effects are therefore ignored for most practical purposes and are otherwise incorporated into the coefficient of discharge. It should be noted that surface tension effects do not appear in problems concerning the single-phase flow of liquids through pipes.

4.3 Scale-up of centrifugal pumps

Show by the method of dimensions that the scale-up of centrifugal pumps can be based on the dimensionless groups

$$\frac{P}{\rho N^3 D^5} = f\left(\frac{Q}{ND^3}, \frac{gH}{N^2 D^2}, \frac{\rho N D^2}{\mu}\right)$$

where P is the power required for pumping, N is the rotational speed and D the diameter of the impeller, H is the head developed, Q is the flow delivered, ρ *is the density and* μ *the viscosity of the fluid, and g is the acceleration due to gravity.*

Solution

The scale-up of centrifugal pumps is a common challenge in the process industries. Experiments may be carried out on a laboratory scale or pilot plant scale as a basis for designing a full-size plant. Alternatively, data may be available from an existing full-size plant and may need conversion to another full-size plant of different capacity and operating conditions. The variables of importance can be combined to give dimensionless groups of power, capacity and head coefficient. This is based on the fact that the power required for pumping is a function of the head developed and volumetric flowrate as well as pump size in terms of impeller diameter and its rotational speed, and the properties of the fluid in terms of density and viscosity. For the purpose of the analysis H is combined with g where the power relationship is assumed to be

$$P = kQ^a (gH)^b N^c D^d \rho^e \mu^f$$

then

$$P \propto Q^a (gH)^b N^c D^d \rho^e \mu^f$$

In terms of the fundamental dimensions M, L and T

$$ML^2T^{-3} = (L^3T^{-1})^a (L^2T^{-2})^b (T^{-1})^c (L)^d (ML^{-3})^e (ML^{-1}T^{-1})^f$$

Equating the indices for M, L and T, respectively

$$1 = e + f$$

$$2 = 3a + 2b + d - 3e - f$$

$$-3 = -a - 2b - c - f$$

The governing variables are flowrate, gravitational acceleration, head (gH) and viscosity. Therefore rearranging in terms of the indices a, b and f

$$e = 1 - f$$

$$c = 3 - a - 2b - f$$

$$\begin{aligned} d &= 2 - 3a - 2b + 3e + f \\ &= 2 - 3a - 2b + 3(1 - f) + f \\ &= 5 - 3a - 2b - 2f \end{aligned}$$

Hence

$$P \propto Q^{a} (gH)^{b} N^{3-a-2b-f} D^{5-3a-2b-2f} \rho^{1-f} \mu^{f}$$

or rearranged into dimensionless groups

$$\frac{P}{\rho N^3 D^5} = f\left(\frac{Q}{ND^3}, \frac{gH}{N^2 D^2}, \frac{\rho N D^2}{\mu}\right)$$

where the power coefficient is a function of the capacity coefficient C_Q, head coefficient C_H, and a form of Reynolds number, respectively.

$$P \propto \rho N^3 D^5 \left(\frac{Q}{ND^3}\right)^{a} \left(\frac{gH}{N^2 D^2}\right)^{b} \left(\frac{\mu}{\rho N D^2}\right)^{f}$$

where

$$C_Q = \frac{Q}{ND^3}$$

(see Problem 9.7, page 249)

and

$$C_H = \frac{gH}{N^2 D^2}$$

4.4 Frictional pressure drop for turbulent flow in pipes

Show, by the Buckingham Π method of dimensional analysis, that the frictional pressure drop Δp_f for a fluid of density ρ and viscosity μ flowing with a velocity v through a circular pipe of inside diameter d, length L and surface roughness ε can be given by the dimensionless groups

$$\frac{\Delta p_f}{\rho v^2} = f\left(\frac{L}{d}, \frac{\varepsilon}{d}, \frac{\mu}{\rho v d}\right)$$

Solution

The Buckingham Π method of dimensional analysis proposed by Edgar Buckingham (1867–1940) in 1915 states that if, in a dimensionally consistent equation, there are n variables in which there are contained m fundamental dimensions, there will be $n - m$ dimensionless groups. Buckingham referred to these groups as Π_1, Π_2, etc, such that

$$f(\Pi_1, \Pi_2, \ldots \Pi_{n-m}) = 0$$

In this case, the functional dependency of frictional pressure drop for the fluid is

$$\Delta p_f = f(L, \varepsilon, \mu, d, \rho, v)$$

Thus, the functional relation is

$$f(\Delta p_f, L, \varepsilon, \mu, d, \rho, v) = 0$$

Any dimensionless group from these variables is therefore of the form

$$\Pi = \Delta p_f^a, L^b, \varepsilon^c, \mu^d, d^e, \rho^f, v^g$$

for which the fundamental dimensions of Π are therefore

$$\Pi = (ML^{-1}T^{-2})^a (L)^b (L)^c (ML^{-1}T^{-1})^d (L)^e (ML^{-3})^f (LT^{-1})^g$$

For the seven variables with three fundamental dimensions (M, L and T) it is expected that there will be four dimensionless groups. The procedure is to select repeating variables by choosing variables equal to the number of fundamental dimensions (three) to provide a succession of Π dimensionless groups which have dimensions which contain amongst them all the fundamental dimensions. It is usually suitable to select ρ, v and d. Thus

$$\Pi_1 = \rho^{a1} v^{b1} d^{c1} \Delta p_f$$

$$\Pi_2 = \rho^{a2} v^{b2} d^{c2} L$$

$$\Pi_3 = \rho^{a3} v^{b3} d^{c3} \varepsilon$$

$$\Pi_4 = \rho^{a4} v^{b4} d^{c4} \mu$$

That is

$$0 = (ML^{-3})^{a1}(LT^{-1})^{b1}(L)^{c1}(ML^{-1}T^{-2})$$

$$0 = (ML^{-3})^{a2}(LT^{-1})^{b2}(L)^{c2}(L)$$

$$0 = (ML^{-3})^{a3}(LT^{-1})^{b3}(L)^{c3}(L)$$

$$0 = (ML^{-3})^{a4}(LT^{-1})^{b4}(L)^{c4}(ML^{-1}T^{-1})$$

Treating each separately using the Rayleigh method gives

$$\Pi_1 = \frac{\Delta p_f}{\rho v^2}$$

$$\Pi_2 = \frac{L}{d}$$

$$\Pi_3 = \frac{\varepsilon}{d}$$

$$\Pi_4 = \frac{\mu}{\rho v d}$$

The four dimensionless groups may be therefore written as

$$\frac{\Delta p_f}{\rho v^2} = f\left(\frac{L}{d}, \frac{\varepsilon}{d}, \frac{\mu}{\rho v d}\right)$$

4.5 Scale model for predicting pressure drop in a pipeline

A half-scale model is used to simulate the flow of a liquid hydrocarbon in a pipeline. Determine the expected pressure drop along the full-scale pipe if the average velocity of the hydrocarbon in the full-scale pipe is expected to be 1.8 ms^{-1}. The model uses water where the pressure drop per metre length is noted to be 4 kNm^{-2}. The respective densities of water and hydrocarbon are 1000 kgm^{-3} and 800 kgm^{-3}, and viscosities are 1×10^{-3} Nsm^{-2} and 9×10^{-4} Nsm^{-2}.

Solution

By dimensional analysis it can be shown that the pressure coefficient is a function of Reynolds number of the fluid in the pipe. That is

$$\frac{\Delta p_f}{\rho v^2} = f\left(\frac{\mu}{\rho v d}\right)$$

Both the model and full-scale pipe are dynamically similar since the resistance to flow is due to viscosity. The Reynolds number for both model and full-scale pipe must therefore be the same. That is

$$\left(\frac{\rho v d}{\mu}\right)_{model} = \left(\frac{\rho v d}{\mu}\right)_{full\ scale}$$

Rearranging in terms of velocity through the model is therefore

$$v_{model} = \frac{\left(\frac{\rho v}{\mu}\right)_{full\ scale}}{\left(\frac{\rho}{\mu}\right)_{model}}\left(\frac{d_{full\ scale}}{d_{model}}\right)$$

$$= \frac{\frac{800 \times 1.8}{9 \times 10^{-4}}}{\frac{1000}{1 \times 10^{-3}}} \times 2$$

$$= 3.2\ \mathrm{ms}^{-1}$$

Since the Reynolds number is the same in both model and full-scale pipe, then so too is the pressure coefficient. That is

$$\left(\frac{\Delta p_f}{\rho v^2}\right)_{model} = \left(\frac{\Delta p_f}{\rho v^2}\right)_{full\ scale}$$

Rearranging, the pressure drop per unit length in the full-scale pipe is therefore

$$\Delta p_f = \Delta p_{f\ model} \frac{(\rho v^2)_{full\ scale}}{(\rho v^2)_{model}}$$

$$= 4000 \times \frac{800 \times 1.8^2}{1000 \times 3.2^2}$$

$$= 1012\ \text{Nm}^{-2}$$

The pressure drop per unit length in the full-scale pipe is expected to be 1.012 kNm^{-2}.

Further problems

(1) State Buckingham's Π theorem.

(2) Outline the procedure used in the Rayleigh method of dimensional analysis.

(3) Show that by the method of dimensional analysis the flow over a rectangular weir can be given by the dimensionless groups

$$\frac{Q}{Bg^{\frac{1}{2}}H^{\frac{3}{2}}}=f\left(\frac{\mu}{\rho g^{\frac{1}{2}}H^{\frac{3}{2}}},\frac{\sigma}{\rho g H^{2}}\right)$$

where B is the breadth of the weir, H is the head, ρ is the density, μ the viscosity and σ the surface tension of the liquid, g is the gravitational acceleration and k is a constant.

(4) Rework the problem of dimensionless groups for flow over a V-notch in which the angle θ is not necessarily the same for all notches to yield

$$\frac{Q}{g^{\frac{1}{2}}H^{\frac{5}{2}}}=f\left(\frac{\mu}{\rho g^{\frac{1}{2}}H^{\frac{3}{2}}},\frac{\sigma}{\rho g H^{2}},\theta\right)$$

where H is the head, ρ is the density, μ the viscosity and σ the surface tension of the liquid, g is the gravitational acceleration and k is a constant.

(5) Show, by the method of dimensions, that the viscous resistance of an oil bearing is dependent on the linear dimensions of the bearing d, the viscosity of the oil μ, speed of rotation N and pressure on the bearing p, and is given by

$$\frac{R}{\mu N d^{3}}=f\left(\frac{p}{\mu N}\right)$$

(6) Show by dimensional analysis that a form of frictional pressure drop for the flow of a fluid through a pipe is

$$\frac{\Delta p_f}{\rho v^{2}}=f\left(\frac{\rho v d}{\mu}\right)$$

where ρ is the density, μ the viscosity and v the velocity of the fluid, and d is the diameter of the pipe.

(7) Show by dimensional analysis that a form of frictional pressure drop for the flow of a fluid through geometrically similar pipes can be expressed by

$$\frac{\Delta p_f}{\rho v^2} = f\left(\frac{\rho v d}{\mu}, \frac{L}{d}\right)$$

where ρ is the density, μ the viscosity and v the velocity of the fluid, and d is the diameter and L the length of pipe.

(8) Show by the method of dynamical analysis that the volumetric flow of a fluid through a central circular orifice located in a pipe can be expressed as

$$Q = kA\sqrt{\frac{\Delta p}{\rho}}$$

where ρ is the density of the fluid, A is the area of the orifice, k is a coefficient which depends on the pipe and orifice dimensions and the Reynolds number, and Δp is the pressure drop across the orifice.

(9) Show that the rate of flow of a liquid of kinematic viscosity ν over a 90° V-notch can be given by

$$\frac{Q}{H^{\frac{5}{2}} g^{\frac{1}{2}}} = f\left(\frac{H^{\frac{3}{2}} g^{\frac{1}{2}}}{\nu}\right)$$

where H is the head and g the acceleration due to gravity.

(10) For liquids flowing along pipelines above a critical velocity, show that pressure drop due to friction per unit length is given by

$$\frac{\Delta p_f}{L} = \frac{\rho v^2}{d} f(Re)$$

where ρ is the density and v the velocity of the fluid, d is the diameter of the pipe and Re is the Reynolds number. Hence show that the friction factor f in the formula $4fLv^2/2gd$ for frictional head loss is a function of the Reynolds number.

Flow measurement by differential head

5

Introduction

The measurement of flow of process fluids is an essential aspect of any process operation, not only for plant control but also for fiscal monitoring purposes. A wide variety of flowmeters is available and it is important to select correctly the flowmeter for a particular application. This requires a knowledge and comprehension of the nature of the fluid to be measured and an understanding of the operating principles of flowmeters.

Before the advent of digital control systems which collect and store flow information, the rate of flow of fluids was usually measured by instruments (flowmeters) using the principle of differential pressure. Today, there is a wide variety of flowmeters available. For single-phase, closed pipe flow, flowmeters are broadly classified into those which are intrusive and those which are non-intrusive to the flow of the fluid. Collectively, the classifications include differential pressure meters, positive displacement meters, mechanical, acoustic and electrically heated meters.

Differential flowmeters indirectly measure velocity, and therefore flow, of a fluid by measuring a differential head. Consequently, they are also known as head or rate meters with the main group of meters being venturi, orifice, nozzle and Pitot tubes — although there are others. Such meters are based on the principle that when liquid flows through a restriction, its velocity increases due to continuity. The increase in kinetic energy evolves from the reduction in pressure through the restriction and it is this relationship which allows a measurement of pressure drop to be related to the velocity and therefore flowrate. The relationship is derived from the Bernoulli equation and is essentially an application of the first law of thermodynamics to flow processes.

Venturi meters are simple fluid flow measuring devices which operate by restricting the flow of fluid, thereby increasing velocity and consequently reducing pressure at the point of restriction. By measuring the differential pressure drop at the point of restriction the flowrate can be readily determined. The device, which has no moving parts, consists of a rapidly tapered section to a throat and gentle downstream expansion section. This design prevents the phenomenon of separation and thus a permanent energy loss. For a well-designed venturi, the discharge coefficient should lie between a value of 0.95 and 0.98. If the discharge coefficient is not known or is unavailable for a particular venturi, a value of 0.97 may be reasonably assumed. Although the venturi meter can be installed in any orientation, care must be taken to use the appropriate equation for flow derived from the Bernoulli equation.

The orifice meter is cheaper to manufacture than the venturi meter, but has the disadvantage of a higher permanent energy loss. The device consists of a plate or diaphragm ideally manufactured from a corrosion and erosion-resistant material, positioned between two flanges in a pipeline. The plate is often centrally drilled, although eccentric and segmental diaphragms are used. The meter operates by increasing the velocity of the fluid as it flows through the restriction and measuring the corresponding differential pressure across the device. There is, however, a region of high turbulence behind the orifice giving rise to a high permanent energy loss and consequently the discharge coefficient is considerably less than that obtained using the venturi meter. A value of 0.6 is frequently used for high flowrates although the coefficient varies with the size of orifice relative to the pipe diameter and rate of flow.

Of the mechanical flowmeters which function on simple fluid flow principles, the rotameter operates using a fixed differential head but a variable flow area. It consists of a vertical tapered tube through which a fluid flows upwards and where the elevation of the float contained within the tapered tube provides an indication of the rate of flow.

5.1 Pitot tube

A Pitot tube is used to determine the velocity of air at a point in a process ventilation duct. A manometer contains a fluid of SG 0.84 and indicates a differential head reading of 30 mm. Determine the local velocity in the duct if the density of air is 1.2 kgm^{-3} and may be assumed constant.

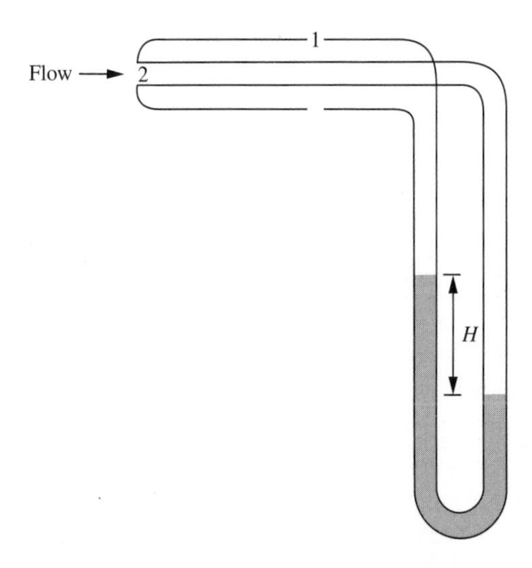

Solution

The Pitot tube, named after the eighteenth century French engineer Henri de Pitot (1695–1771) who invented it, is a device used to measure the local or point velocity of a fluid in a pipe or duct. It can also be used to determine flowrate by measurements of local velocities in the cross-section of the pipe or duct and is particularly useful therefore where the velocity profile is irregular. The device operates by measuring the difference between the impact and static pressures in the fluid and usually consists of two concentric tubes arranged in parallel; one with an opening in the direction of flow, the other perpendicular to the flow. Applying the Bernoulli equation, where there is virtually no head loss due to the proximity of the two points

$$\frac{p_1}{\rho g} + \frac{v_1^2}{2g} = \frac{p_2}{\rho g}$$

The tube is also assumed to be small with respect to pipe size; otherwise, corrections for disturbances and flow area reduction are required. Since the fluid at point 2 is stationary, p_2 therefore corresponds to the impact pressure. Rearranging

$$v_1 = \sqrt{\frac{2(p_2 - p_1)}{\rho}}$$

For the differential manometer, the density of the manometric fluid, ρ_m, is considerably greater than the density of air, ρ. The approximate differential pressure is thus

$$\begin{aligned} p_2 - p_1 &= \rho_m gH \\ &= 840 \times g \times 0.03 \\ &= 247 \text{ Nm}^{-2} \end{aligned}$$

The velocity of the air in the duct is therefore

$$\begin{aligned} v_1 &= \sqrt{\frac{2 \times 247}{1.2}} \\ &= 20.29 \text{ ms}^{-1} \end{aligned}$$

The velocity is found to be 20.29 ms^{-1}. Note that the density of the air can be readily determined applying the ideal gas law for a known barometric pressure and temperature. It may also be noted that the actual velocity measured by the Pitot is given by

$$v = C\sqrt{2gH}$$

where C is a coefficient which is approximately unity for large pipes and smooth Pitots but is appreciably less for low Reynolds number flow.

5.2 Pitot traverse

Determine the rate of flow and average velocity of air in a process vent pipe of 50 cm diameter for which local readings of velocity from a Pitot tube are recorded below.

Radius r (m)	*0*	*0.05*	*0.10*	*0.15*	*0.20*	*0.225*	*0.25*
Velocity v (ms^{-1})	*19.0*	*18.6*	*17.7*	*16.3*	*14.2*	*12.9*	*0*

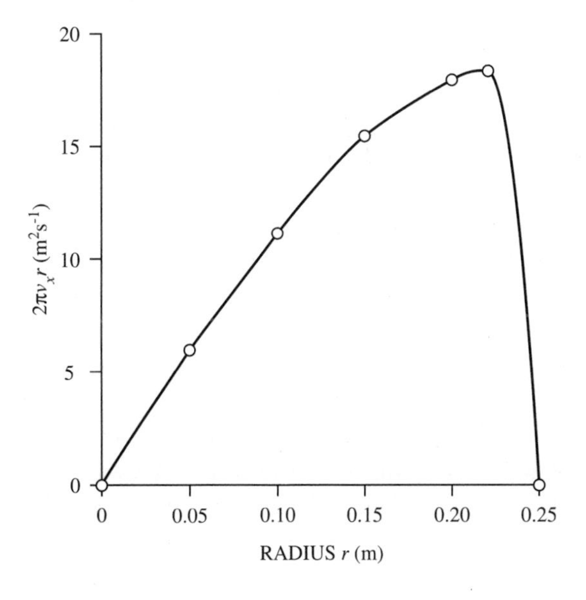

Solution

The distribution of velocity across the process vent pipe using a Pitot tube, known as a Pitot traverse, can be used to determine flowrate either numerically or by graphical integration. The elemental flowrate for the device at a radius r, recording the local velocity v_x, is therefore

$$dQ = v_x 2\pi r\, dr$$

The total flow may therefore be determined graphically from a plot of $2\pi v_x r$ versus r. Thus

Radius r (m)	0	0.05	0.10	0.15	0.20	0.225	0.25
Velocity v (ms^{-1})	19.0	18.6	17.7	16.3	14.2	12.9	0
$2\pi v_x r$ (m^2s^{-1})	0	5.8	11.1	15.4	17.8	18.2	0

From the plot, the area under the curve is found to be 2.74 m^3s^{-1}. The average velocity is found from the total flow across the flow area. That is

$$v = \frac{4Q}{\pi d^2}$$

$$= \frac{4 \times 2.74}{\pi \times 0.5^2}$$

$$= 13.94 \text{ ms}^{-1}$$

The average velocity is found to be 13.94 ms^{-1}. Note that the average velocity is not the sum of the velocities reported divided by the number of readings.

This procedure is applicable only to symmetrical velocity distributions. For unsymmetrical flow or flow in non-circular ducts, the procedure can be modified to evaluate the total flow over the flow section. For rectangular ducts this involves dividing the cross-section into regular sized squares and measuring the local velocities at these points. Alternatively, log-linear for circular cross-sections or log-Tchebychev positions for both circular and rectangular cross-sections can reduce the computations involved but require accurate positioning. Numerical velocity-area integration techniques are, however, now preferred to graphical techniques where the accuracy of the final result depends on the number of measurements made. Ideally, for circular cross-sections there should be about 36 points with 6 on each equi-spaced radii and not less than 12 with 3 on each of 4 equi-spaced radii.

5.3 Horizontal venturi meter

A horizontal venturi meter with a discharge coefficient of 0.96 is to be used to measure the flowrate of water up to 0.025 $m^3 s^{-1}$ in a pipeline of internal diameter 100 mm. The meter is connected to a differential manometer containing mercury of SG 13.6. If the maximum allowable difference in mercury levels is 80 cm, determine the diameter of the throat and the shortest possible overall length of the meter.

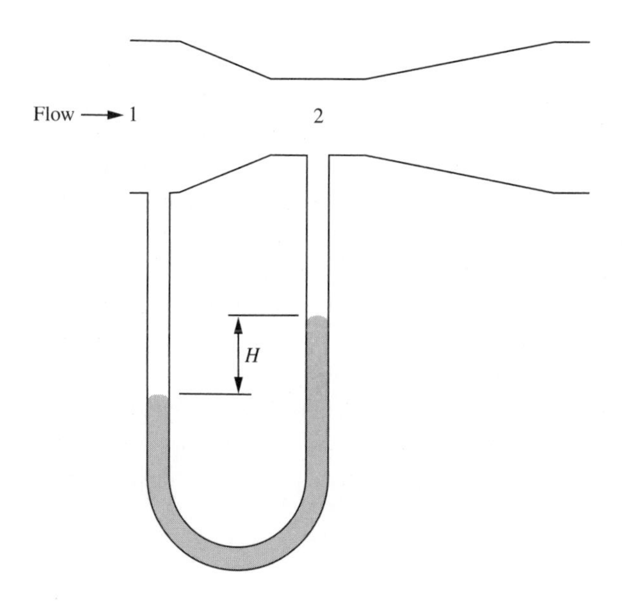

Solution

This fluid flow measuring device, first named after the Italian physicist Giovanni Battista Venturi (1746–1822) by Clemens Hershels in 1886, consists of a tapered tube which constricts flow so that the differential pressure produced by the flowing fluid through the throat indirectly gives a measure of flowrate. The rate of flow can be determined by applying the Bernoulli equation at some point upstream of the venturi (point 1) and at the throat (point 2). For a horizontal venturi

$$\frac{p_1}{\rho g} + \frac{v_1^2}{2g} = \frac{p_2}{\rho g} + \frac{v_2^2}{2g}$$

From continuity for an incompressible fluid

$$a_1 v_1 = a_2 v_2$$

Substituting for v_2 and rearranging

$$\frac{p_1 - p_2}{\rho g} = \frac{v_1^2}{2g}\left(\left(\frac{a_1}{a_2}\right)^2 - 1\right)$$

Rearranging, the velocity in the pipe v_1 is therefore

$$v_1 = \sqrt{\frac{2(p_1 - p_2)}{\rho\left(\left(\frac{a_1}{a_2}\right)^2 - 1\right)}}$$

The actual flow through the venturi incorporates a coefficient of discharge C_d to allow for frictional effects and is defined as the ratio of actual to theoretical flow. Therefore

$$Q = C_d a v_1$$

$$= C_d a \sqrt{\frac{2(p_1 - p_2)}{\rho(\beta^4 - 1)}}$$

where β is the ratio of pipe diameter to throat diameter and a is the flow area of the pipe. For a pipe diameter of 0.1 m the pipe flow area is

$$a = \frac{\pi \times 0.1^2}{4}$$

$$= 7.85 \times 10^{-3}\ \text{m}^2$$

Rearranging, the flow equation in terms of β is

$$\beta = \left(\frac{2(p_1 - p_2)}{\rho\left(\frac{Q}{C_d a}\right)^2} + 1\right)^{\frac{1}{4}}$$

For the mercury-filled differential manometer the differential pressure is

$$
\begin{aligned}
p_1 - p_2 &= (\rho_{Hg} - \rho)gH \\
&= (13{,}600 - 1000) \times g \times 0.8 \\
&= 98{,}885 \text{ Nm}^{-2}
\end{aligned}
$$

β is therefore

$$
\begin{aligned}
\beta &= \left(\frac{2 \times 98{,}885}{1000 \times \left(\frac{0.025}{0.96 \times 7.85 \times 10^{-3}} \right)^2} + 1 \right)^{\frac{1}{4}} \\
&= 2.087 \\
&= \frac{d_1}{d_2}
\end{aligned}
$$

The diameter of the throat is therefore

$$
\begin{aligned}
d_2 &= \frac{d_1}{\beta} \\
&= \frac{0.1}{2.087} \\
&= 0.0479 \text{ m}
\end{aligned}
$$

That is, the throat diameter is found to be 48 mm. The dimensions of the venturi are important to minimize permanent energy losses. The recommended dimensions of a venturi suggest an inlet entrance cone to have an angle of 15°–20° with an exit cone angle of 5°–7.5°. The throat length is 0.25 to 0.5 pipe-diameters with tapping points located between 0.25 and 0.75 pipe-diameters upstream. In this case, the shortest possible overall length of venturi is therefore an entrance cone of 7.1 cm length (20°), a throat of 2.5 cm (0.25 pipe-diameters) and an exit cone of 19.7 cm (7.5°) giving an overall length of 29.3 cm.

5.4 Orifice and venturi meters in parallel

An orifice plate meter and a venturi meter are connected in parallel in a horizontal pipe of inside diameter 50 mm. The orifice has a throat diameter of 25 mm and discharge coefficient of 0.65 while the venturi has a throat diameter of 38 mm and discharge coefficient of 0.95. Determine the proportion of flow through either meter.

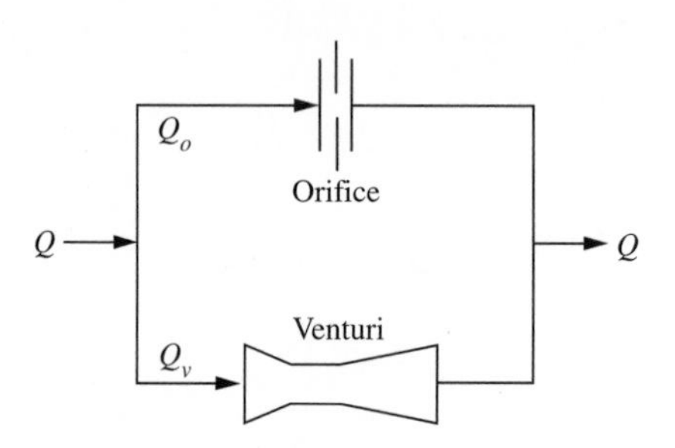

Solution

It is assumed that the liquid flows along a frictionless horizontal pipe and that losses due to fittings can be neglected. For liquid to flow across both instruments, the pressure drop across them must be the same. Due to the different characteristics of these two meters, the flow through the orifice should be less than that through the venturi since the overall pressure drop for a given flow is higher through an orifice than a venturi. Thus for both meters, the flowrate is

$$Q_o = C_{d(o)} a \sqrt{\frac{2\Delta p}{\rho\left(\left(\frac{a}{a_o}\right)^2 - 1\right)}}$$

(see Problem 5.3, page 119)

$$Q_v = C_{d(v)} a \sqrt{\frac{2\Delta p}{\rho\left(\left(\frac{a}{a_v}\right)^2 - 1\right)}}$$

where the ratio of pipe to orifice area in terms of diameter is

$$\frac{a}{a_o} = \left(\frac{d}{d_o}\right)^2$$

and ratio of pipe to venturi throat area in terms of diameter is

$$\frac{a}{a_v} = \left(\frac{d}{d_v}\right)^2$$

Since the differential pressure across both meters is the same, then

$$\left(\frac{Q_o}{C_{d(o)}}\right)^2\left(\left(\frac{d}{d_o}\right)^4 - 1\right) = \left(\frac{Q_v}{C_{d(v)}}\right)^2\left(\left(\frac{d}{d_v}\right)^4 - 1\right)$$

That is

$$\left(\frac{Q_o}{0.65}\right)^2\left(\left(\frac{0.05}{0.025}\right)^4 - 1\right) = \left(\frac{Q_v}{0.95}\right)^2\left(\left(\frac{0.05}{0.038}\right)^4 - 1\right)$$

which reduces to

$$Q_v = 4Q_o$$

From continuity, the total flow Q is the sum of the flow through the meters. That is

$$\begin{aligned} Q &= Q_o + Q_v \\ &= Q_o + 4Q_v \\ &= 5Q_o \end{aligned}$$

Thus 20% of the flow passes through the orifice meter while 80% of the flow passes through the venturi. Note that for the same throat size, the difference in volumetric flowrate is entirely due to the permanent frictional losses imposed by both meters.

5.5 Venturi meter calibration by tracer dilution

A venturi meter with a throat diameter of 6 cm is used to measure the flowrate of water of density 1000 kgm^{-3} along a horizontal pipeline with an inside diameter of 100 mm. The flowmeter is calibrated using a solution of salt water with a salt concentration of 20 gl^{-1} and added continuously upstream of the meter at a rate of 3 litres per minute. Determine the discharge coefficient of the meter if a pressure drop across the throat of 3.7 kNm^{-2} is recorded and a diluted salt concentration of 0.126 gl^{-1} is found by analysis downstream.

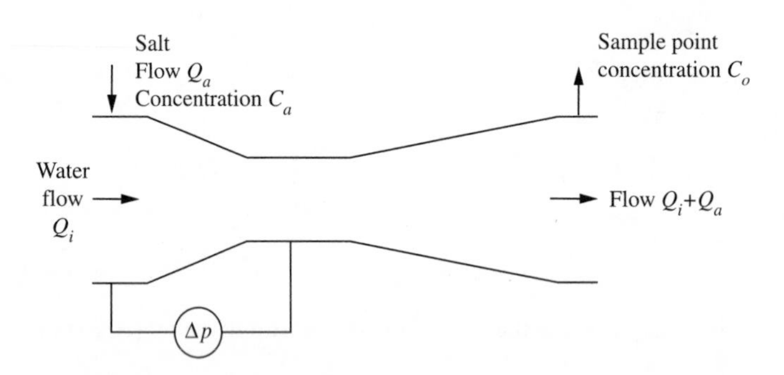

Solution

The calibration of flowmeters by the method of tracer dilution involves the addition of extraneous material whose presence can be quantitatively determined by an analytical technique. The extraneous material may or may not be already present as an impurity. In this case, the flowrate of water through the meter is related to diluted salt concentration. From a material balance on the salt

$$Q_a C_a = (Q_i + Q_a) C_o$$

where Q_i and Q_a are the rates of flow of water and salt solution, and C_a and C_o are the added upstream and diluted downstream salt concentrations. Rearranging, the flowrate of water is therefore

$$Q_i = Q_a \left(\frac{C_a - C_o}{C_o} \right)$$

$$= \frac{0.003}{60} \times \left(\frac{20 - 0.126}{0.126} \right)$$

$$= 7.89 \times 10^{-3} \text{ m}^3\text{s}^{-1}$$

The total flow Q through the meter during the calibration is therefore

$$Q = Q_i + Q_a$$

$$= 7.89 \times 10^{-3} + 5 \times 10^{-5}$$

$$= 7.895 \times 10^{-3}\ \text{m}^3\text{s}^{-1}$$

The rate of flow through a horizontal venturi meter is given by

$$Q = C_d a \sqrt{\frac{2\Delta p}{\rho(\beta^4 - 1)}}$$

where β is the ratio of pipe to throat diameter

$$\beta = \frac{10}{6}$$

$$= 1.67$$

Rearranging in terms of discharge coefficient

$$C_d = \frac{Q}{a\sqrt{\dfrac{2\Delta p}{\rho(\beta^4 - 1)}}}$$

$$= \frac{7.895 \times 10^{-3}}{\dfrac{\pi \times 0.1^2}{4} \times \sqrt{\dfrac{2 \times 3700}{1000 \times (1.67^4 - 1)}}}$$

$$= 0.962$$

The coefficient of discharge is found to be 0.962.

5.6 Differential pressure across a vertical venturi meter

A process liquid of density 850 kgm^{-3} flows upward at a rate of 0.056 m^{3}s^{-1} through a vertical venturi meter which has an inlet diameter of 200 mm and throat diameter of 100 mm, with discharge coefficient of 0.98. Determine the difference in reading of two pressure gauges located at the respective tapping points a vertical distance of 30 cm apart.

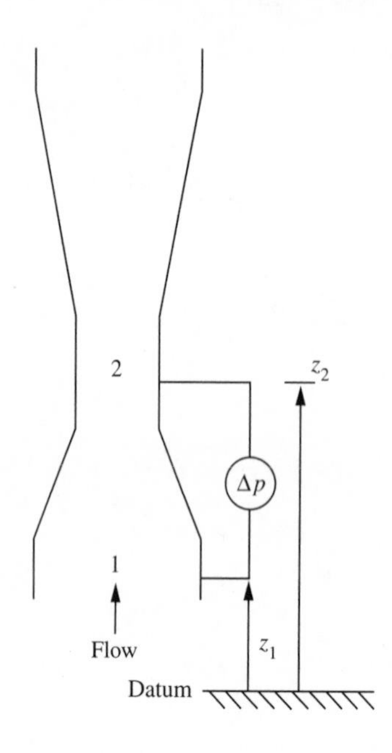

Solution

Applying the Bernoulli equation between the upstream position (1) and the throat (2)

$$\frac{p_1}{\rho g} + \frac{v_1^2}{2g} + z_1 = \frac{p_2}{\rho g} + \frac{v_2^2}{2g} + z_2$$

From continuity for an incompressible fluid

$$a_1 v_1 = a_2 v_2$$

Rearranging in terms of differential pressure

$$p_1 - p_2 = \rho g\left(\frac{v_1^2}{2g}\left(\left(\frac{a_1}{a_2}\right)^2 - 1\right) + z_2 - z_1\right)$$

The velocity is related to flowrate by

$$Q = C_d a v_1$$

and for the circular cross-section

$$\left(\frac{a_1}{a_2}\right)^2 = \left(\frac{d_1}{d_2}\right)^4$$

therefore

$$p_1 - p_2 = \frac{\rho}{2}\left(\left(\frac{Q}{C_d a}\right)^2\left(\left(\frac{d_1}{d_2}\right)^4 - 1\right) + 2g(z_2 - z_1)\right)$$

$$= \frac{850}{2} \times \left(\left(\frac{0.056}{0.98 \times \frac{\pi \times 0.2^2}{4}}\right)^2 \times \left(\left(\frac{0.2}{0.1}\right)^4 - 1\right) + 2 \times g \times 0.3\right)$$

$$= 23{,}594 \text{ Nm}^{-2}$$

The differential pressure is found to be 23.6 kNm^{-2}. Note that a differential manometer would not record a difference in level of manometric fluid when there is no flow through the venturi whereas two independent pressure gauges would record a static pressure difference equal to

$$p_1 - p_2 = \rho g(z_2 - z_1)$$

5.7 Flow measurement by orifice meter in a vertical pipe

Oil of density 860 kgm^{-3} flows up a vertical pipe section of diameter 225 mm. A manometer filled with fluid of density 1075 kgm^{-3} is used to measure the pressure drop across an orifice plate with a throat diameter of 75 mm. Determine the flowrate of oil if the deflection of the manometer fluid is 0.5 m. Assume a discharge coefficient of 0.659 for the orifice.

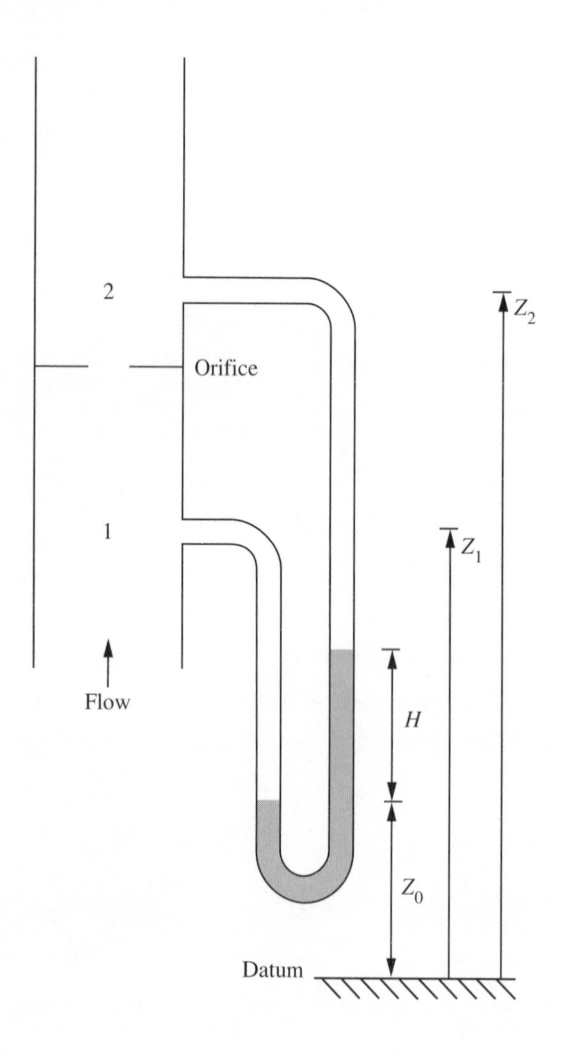

Solution

Applying the Bernoulli equation between points 1 and 2

$$\frac{p_1}{\rho g} + \frac{v_1^2}{2g} + z_1 = \frac{p_2}{\rho g} + \frac{v_2^2}{2g} + z_2$$

From continuity for an incompressible fluid

$$a_1 v_1 = a_2 v_2$$

Therefore substituting for v_2 and rearranging

$$p_1 - p_2 + \rho g(z_1 - z_2) = \frac{\rho v_1^2}{2}\left(\left(\frac{a_1}{a_2}\right)^2 - 1\right)$$

For the manometer

$$p_1 - p_2 = \rho g(z_2 - z_1) + (\rho_m - \rho)gH$$

The theoretical velocity through the pipe is therefore

$$v_1 = \sqrt{\frac{2gH(\rho_m - \rho)}{\rho(\beta^4 - 1)}}$$

$$= \sqrt{\frac{2 \times g \times 0.5 \times (1075 - 860)}{860 \times \left(\left(\frac{0.225}{0.075}\right)^4 - 1\right)}}$$

$$= 0.523 \text{ ms}^{-1}$$

Note that this expression does not contain terms in z. The velocity and therefore flowrate is independent of the orientation of the pipe. The actual flowrate is then

$$Q = C_d a_1 v_1$$

$$= 0.659 \times \frac{\pi \times 0.225^2}{4} \times 0.523$$

$$= 0.014 \text{ m}^3\text{s}^{-1}$$

The rate of flow is found to be 0.014 m^3s^{-1}.

5.8 Variable area flowmeter

A rotameter used to measure the flow of water consists of a float with a mass of 30 g set in a tapered glass tube 20 cm in length. The tube has an internal diameter of 22 mm at its base and 30 mm at its top end. Determine the rate of flow when the float is at mid-height in the tube. The rotameter has a coefficient of discharge of 0.6 and the density of the float is 5100 kgm^{-3}.

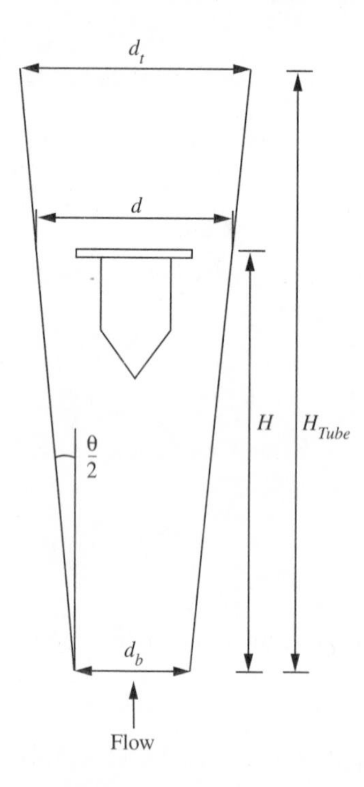

Solution

Often referred to as a rotameter, the variable area flowmeter is used to measure the rate of flow of a fluid by virtue of the elevation of a solid float within a vertical tapered tube. The tube is transparent and is usually made of glass while the float is made of metal, ceramic or plastic and is usually 'bomb' shaped, in that it has a cylindrical body with a cone-shaped bottom and short flat top piece. Some floats are grooved which encourages the float to spin thereby improving stability due to a gyroscopic effect. This device has effectively a fixed permanent pressure drop and a variable flow area. As the fluid flowrate is increased, the float moves up the tube until an equilibrium position of forces is reached.

From Archimedes' principle, this is a position where the upthrust is equal to the weight of fluid displaced. That is

$$\Delta p A_f = \rho_f V_f g - \rho V_f g$$

where A_f is the cross-sectional area of float, ρ_f is the density of the float, ρ is the density of fluid and V_f is the volume of float. Rearranging in terms of Δp then

$$\Delta p = \frac{V_f g(\rho_f - \rho)}{A_f}$$

where Δp is related to flowrate by the Bernoulli equation. At the elevation of the float

$$\frac{p + \Delta p}{\rho g} + \frac{v_t^2}{2g} = \frac{p}{\rho g} + \frac{v_a^2}{2g}$$

where v_t and v_a are the velocities of the fluid in the tube and annulus, respectively. Therefore

$$\Delta p = \frac{\rho}{2}(v_a^2 - v_t^2)$$

Since the velocity of the fluid through the annulus is significantly greater than that in the tube, this approximates to

$$\Delta p = \frac{\rho v_a^2}{2}$$

Combining both equations for Δp, the velocity of the fluid through the annulus is therefore

$$v_a = \sqrt{\frac{2V_f g(\rho_f - \rho)}{\rho A_f}}$$

Introducing a discharge coefficient C_d to allow for losses due to friction, the actual mass flowrate is therefore given by

$$\begin{aligned} m &= C_d \rho a_a v_a \\ &= C_d \rho a_a \sqrt{\frac{2V_f g(\rho_f - \rho)}{\rho A_f}} \\ &= C_d a_a \sqrt{\frac{2V_f g\rho(\rho_f - \rho)}{A_f}} \end{aligned}$$

If the density of the fluid is not influenced by temperature or composition, then the square root term remains constant. Further, if the discharge coefficient does not vary greatly, then there is an almost linear variation between mass flowrate and the area of the annulus. To relate the area of the annulus to the float position above the bottom of the tube, the float is assumed to sit perfectly in the bottom of the tube of internal diameter d_b when there is no flow. From geometry, the internal diameter of the tube d for any float position above the bottom H, for an angle of taper θ, is therefore

$$d = d_b + 2H \tan\frac{\theta}{2}$$

for which the area of the annulus, a_a, around the float is

$$\begin{aligned} a_a &= \frac{\pi}{4}(d^2 - d_b^2) \\ &= \frac{\pi}{4}\left(\left(d_b + 2H \tan\frac{\theta}{2}\right)^2 - d_b^2\right) \\ &= \frac{\pi}{4}\left(d_b^2 + 4d_bH \tan\frac{\theta}{2} + 4H^2\left(\tan\left(\frac{\theta}{2}\right)\right)^2 - d_b^2\right) \end{aligned}$$

This approximates to

$$a_a = \pi d_b H \tan\frac{\theta}{2}$$

The mass flowrate is therefore

$$\begin{aligned} m &= C_d \pi d_b H \tan\frac{\theta}{2}\sqrt{\frac{2V_f g\rho(\rho_f - \rho)}{\frac{\pi d_b^2}{4}}} \\ &= C_d H \tan\frac{\theta}{2}\sqrt{8V_f g\rho(\rho_f - \rho)\pi} \end{aligned}$$

That is, the mass flowrate is approximately linear with float position in the tube. In practice, a scale is marked on the tube and the rotameter is supplied with a calibration curve for a particular fluid, temperature and float. In this case, the float has a volume

$$V_f = \frac{m_f}{\rho_f}$$

$$= \frac{0.03}{5100}$$

$$= 5.88 \times 10^{-6}\,\text{m}^3$$

The half-angle of taper is related to top and bottom diameters separated by a height H_{Tube}

$$\tan\frac{\theta}{2} = \frac{d_t - d_b}{2H_{Tube}}$$

$$= \frac{0.03 - 0.22}{2 \times 0.2}$$

$$= 0.02$$

The mass flowrate through the rotameter for a float at the mid-point in the tube (10 cm) is therefore

$$m = 0.6 \times 0.1 \times 0.02 \times \sqrt{8 \times 5.88 \times 10^{-6} \times g \times 1000 \times (5100 - 1000) \times \pi}$$

$$= 0.0925\ \text{kgs}^{-1}$$

corresponding to a volumetric flow of 5.55 litres per minute.

5.9 Rotameter calibration by venturi meter

A horizontal venturi meter with a throat diameter of 2.5 cm is used to calibrate a vertical rotameter in a pipeline carrying water. The pipeline has an inside diameter of 5 cm. The rotameter has a cone angle of 2° and uses a float with a volume of 100 cm^3 and density 8000 kgm^{-3}. Determine the flowrate of water and the discharge coefficient for the rotameter if a differential pressure reading of 1.2 kNm^{-2} is recorded for the venturi when the float is at an elevation of 15 cm. The discharge coefficient for the venturi is 0.97.

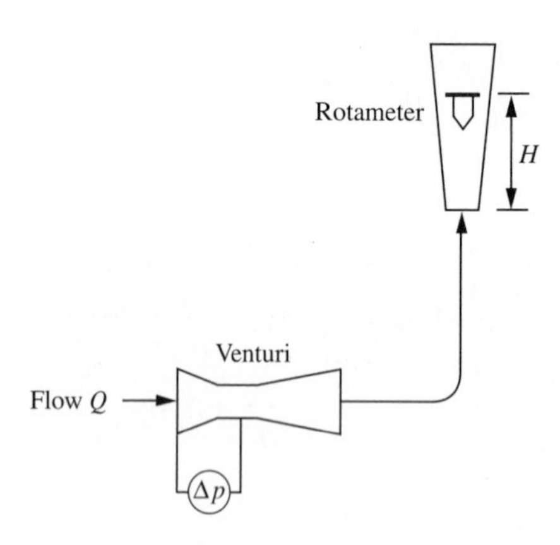

Solution

The flow of water through the horizontal venturi is

$$Q = C_d a \sqrt{\frac{2\Delta p}{\rho(\beta^4 - 1)}}$$

where β is the ratio of pipe to throat diameter

$$\beta = \frac{d_1}{d_2}$$

$$= \frac{0.05}{0.025}$$

$$= 2$$

Therefore

$$Q = 0.97 \times \frac{\pi \times 0.05^2}{4} \times \sqrt{\frac{2 \times 1200}{1000 \times (2^4 - 1)}}$$

$$= 7.62 \times 10^{-4} \text{ m}^3\text{s}^{-1}$$

The flow through the rotameter is given by

$$Q = \frac{C_d H}{\rho} \tan\frac{\theta}{2} \sqrt{8 V_f g \rho(\rho_f - \rho)\pi}$$

Rearranging, the discharge coefficient is therefore

$$C_d = \frac{Q}{\frac{H}{\rho} \tan\frac{\theta}{2} \sqrt{8 V_f g \rho(\rho_f - \rho)\pi}}$$

$$= \frac{7.62 \times 10^{-4}}{\frac{0.15}{1000} \times \tan\frac{2°}{2} \times \sqrt{8 \times 1 \times 10^{-4} \times g \times 1000 \times (8000 - 1000) \times \pi}}$$

$$= 0.70$$

The coefficient for the rotameter is found to be 0.70.

Further problems

(1) Water flows down a vertical tapering pipe 2 m long. The top of the pipe has a diameter of 10 cm and the diameter of the bottom of the pipe is 5 cm. Determine the difference of pressure between the top and the bottom ends of the pipe if the flow rate is 1 m^3min^{-1}.

Answer: 1766 Nm^{-2}

(2) The velocity of water in a pipe with a bore of 250 mm is measured with a Pitot tube. The difference in head at the centre of the pipe is found to be 10 cm of water. Determine the flowrate of water per minute if the average velocity of water is two-thirds the velocity at the centre. The coefficient of the Pitot tube may be taken as unity.

Answer: 0.0458 m^3s^{-1}

(3) A liquid of density 800 kgm^{-3} flows through a horizontal pipe with an inside diameter of 150 mm under a pressure of 400 kNm^{-2}. Assuming no losses, determine the flow when the pressure at a 75 mm diameter reduction is 200 kNm^{-2}.

Answer: 0.102 m^3s^{-1}

(4) Starting with the Bernoulli equation, show that the volumetric flowrate of an incompressible fluid through a horizontal venturi meter where the throat and upstream positions are connected by a U-tube containing a fluid of density ρ_m can be given by

$$Q = C_d a \sqrt{\frac{2gH\left(\frac{\rho_m}{\rho} - 1\right)}{(\beta^4 - 1)}}$$

where C_d is the discharge coefficient, a is the pipe flow area, ρ is the fluid density, β is the ratio of the pipe diameter to throat diameter, g is the gravitational acceleration and H is the difference in levels of the fluid in the U-tube.

(5) A venturi meter of inlet diameter 10 cm and throat diameter 5 cm is installed in a vertical pipe through which water flows upwards. The meter is calibrated where it is noted that for a flow rate of 11.5 litres per second, the difference in readings of the pressure gauges connected to the inlet and throat is

20 kNm^{-2}. If the difference in the height of the two tapping points is 30 cm, determine the discharge coefficient for the venturi meter.

Answer: 0.971

(6) A venturi meter is installed in a horizontal pipe with an inside diameter of 100 mm carrying an organic solvent of density 1200 kgm^{-3}. The only equipment available for measuring pressure differences is a manometer containing mercury of SG 13.6, with a maximum allowable difference of levels of 80 cm. Assuming a discharge coefficient is 0.97, determine the maximum flowrate of solvent which can be measured.

Answer: 0.0251 m^3s^{-1}

(7) The flow in a pipeline of diameter d is measured by an orifice meter with an orifice diameter of $0.5d$. The discharge coefficient is 0.62. Calculate the throat diameter of a horizontal venturi meter which would give the same pressure difference at the same rate of flow in the pipe. The discharge coefficient for the venturi meter may be taken as 0.96.

Answer: $0.405d$

(8) Discuss briefly the merits of both the orifice plate meter and the venturi meter as differential pressure meters.

(9) Describe the types of orifice plate which are available and their application.

(10) A horizontal venturi meter with a 5 cm diameter throat is used to measure the flow of slightly salt water in a pipe of inside diameter 10 cm. The meter is calibrated by adding 2 litres per minute of 0.5 molar sodium chloride solution upstream of the meter and analysing a sample of water downstream from the meter. Before the addition of the salt solution, a 1 litre sample required 10 cm^3 of 0.1 molar silver nitrate solution in a titration. After addition, a 1 litre sample required 24.4 cm^3 of 0.1 molar silver nitrate. Determine the discharge coefficient of the meter for a differential pressure across the venturi of 17.4 kNm^{-2}.

Answer: 0.96

(11) Identify suitable types of tracer substances with their appropriate detection systems which can be used to calibrate flowmeters.

(12) An organic liquid of density 980 kgm^{-3} flows with an average velocity of 3 ms^{-1} along a 75 mm diameter horizontal pipe. A restriction is placed in the pipe in which there is a 50 mm diameter opening for the water to flow. The pressure of the water in the 75 mm pipe is 150 kNm^{-2}. Determine the pressure at the restriction.

Answer: 144.2 kNm^{-2}

(13) The difference in head registered between two limbs of a mercury gauge, with water above the mercury and connected to a horizontal venturi meter, is 20 cm. If the venturi has a pipe and throat diameter of 15 cm and 7.5 cm, respectively, determine the discharge through the meter assuming a coefficient of discharge of 0.97.

Answer: 0.031 m^3s^{-1}

(14) The position of a float, of density 3000 kgm^{-3}, in a rotameter for a particular flow is 15 cm above the bottom of the tapered tube. If the float is replaced with an identical size of float but with a density of 2500 kgm^{-3}, then the position of the float in the tube is increased by 2.5 cm. Determine the density of the fluid.

Answer: 1115 kgm^{-3}

(15) Highlight the advantages and disadvantages of using rotameters for the measurement of fluid flow.

(16) A uniform pipeline of circular cross-section is to carry water at a rate of 0.5 m^3s^{-1}. If an orifice plate is to be used to monitor the flowrate in the line, determine the differential pressure reading across the plate at the design flowrate. The pipe has an internal diameter of 45 cm and the orifice has a concentric hole of 30 cm. If the differential pressure is to be measured using a mercury-filled manometer, determine the difference in level between the two legs. The coefficient of discharge for the plate may be taken as 0.6.

Answer: 55.8 kNm^{-2}, 45.1 cm

(17) An undergraduate laboratory experiment involves a venturi meter which is used to measure the rate of flow of water through a horizontal pipe with an internal diameter of 5 cm. A calibration is provided for the meter and is given by

$$Q = 0.096\sqrt{H}$$

where Q is the flowrate in litres per second and H is the manometric head in a water/mercury manometer measured in millimetres. Determine the throat diameter of the venturi if the coefficient of discharge may be assumed to be 0.97.

Answer: 16.9 mm

(18) A Pitot tube is used to measure the velocity of water at a point in an open channel. If the Pitot is connected to an open air-filled inverted manometer, determine the velocity of the water for a difference in liquid levels in the manometer legs of 10 mm.

Answer: 0.44 ms^{-1}

(19) A Pitot traverse is used to record the local velocities of air across the entrance of a circular duct of diameter 91 cm. The Pitot is connected to a water-filled differential manometer and measurements of the difference in levels between the legs are given below:

Distance from pipe wall, cm	0	5	10	20	30	40	45.5
Difference in levels, mm	0	7	10	12	14	15	15

If the air has a static pressure of 101.3 kNm^{-2} and temperature of 20°C, determine the rate of flow and the average velocity. The mean molecular mass of air is 29 $kgkmol^{-1}$ and the Universal gas constant is 8.314 $kJkmol^{-1}K^{-1}$.

Answer: 7.5 m^3s^{-1}, 11.5 ms^{-1}

(20) Outline the advantages and disadvantages of Pitot tubes for the measurement of fluid velocity and flow.

(21) A venturi meter measures the flow of water in a 100 mm inside diameter horizontal pipe. The difference in head between the entrance and the throat of the meter is measured by a U-tube, containing mercury (SG 13.6) with the space above the mercury in each limb being filled with water. Determine the diameter of the throat of the meter such that the difference in the levels of mercury shall not exceed 300 mm when the quantity of water flowing in the pipe is 10 kgs^{-1}. Assume the discharge coefficient is 0.97.

Answer: 39 mm

Tank drainage and variable head flow

6

Introduction

The rate at which fluids freely discharge from tanks and vessels through orifices and connecting pipes is dependent on the pressure or head within the tank and on frictional resistance. The shape, size and form of the orifice through which the fluid discharges, and the length and diameter of pipe and fluid properties, also influence the rate of discharge due to the effects of frictional resistance. As the fluid passes through the orifice issuing as a free jet, it contracts in area, thus further reducing the rate of discharge. The contraction is caused by the liquid in the vessel in the vicinity of the orifice having a motion perpendicular to that of the jet and exerting a lateral force. The section of the jet at which the streamlines become parallel is known as the *vena contracta*, and there is no further contraction beyond this point. Owing to the reduction in both velocity and flow area of the jet, the actual rate of flow is much less than the theoretical prediction: the relation between them is known as the coefficient of discharge.

The simplest method of determining the coefficient of discharge is to measure the quantity of fluid (liquid) discharged for a constant head and divide by the theoretical discharge. In terms of energy conversion for a freely draining tank, the potential energy is converted to kinetic energy or, in the head form, the static head is thus converted to velocity head. It is therefore deduced that the theoretical velocity of the jet is proportional to the square root of the head, described in what is called the Torricelli equation after the Italian mathematician Evangelista Torricelli (1608–1647). The difference between the actual and theoretical velocities, known as the coefficient of velocity, is due to friction at the orifice and is small for sharp-edged orifices. The coefficient of velocity varies depending on the size and shape of the orifice as well as the applied head. Nevertheless, typical values are in the order of 0.97 and may be found experimentally for a vertically-mounted orifice in the side of a tank by measuring the horizontal and vertical co-ordinates of the issuing jet's trajectory.

For tank drainage problems which do not involve simultaneous inflow to maintain a constant head above the orifice, there is a continuous loss of capacity and consequently of head. The rate of discharge is therefore not constant but variable, with the head being dependent on the geometry of the tank. The time taken for tank drainage can be considered by applying an unsteady state mass balance over the tank and equating the rate of discharge with the loss of capacity from the tank. Analytical solutions can be deduced for most tank configurations in terms of tank geometry, for simultaneous inflow, flow into adjacent vessels through openings, submerged orifices (sluices), and for flow through connecting pipes with laminar flow. Problems in tank drainage which involve the flow through connecting pipes with turbulent flow are not, however, mathematically straightforward.

In simple treatments of tank drainage problems, it is necessary to employ pertinent assumptions. These include a well-ventilated tank where the applied pressure (usually atmospheric) is the same at both the free surface of the liquid in the tank and at the jet, where there is no free vortex formation and where the discharge coefficient is constant. In practice, however, each of these has an influence and should be allowed for where appropriate by more detailed calculations.

6.1 Orifice flow under constant head

A water cooling tower receives 5000 m^3 of water per day. If a water distribution system is to be designed to cover the area of the tower using 1 cm diameter orifices in the base of the ducting in which the water will be at a depth of 20 cm, determine the number of orifices required. A discharge coefficient of 0.6 is assumed for the orifices.

Solution

To determine the rate at which the liquid discharges through the orifices, it is necessary to determine the velocity of the liquid in the free jets. Assuming that there is no pressure change across the free jet, the static head is directly converted to velocity head as

$$H = \frac{v^2}{2g}$$

Rearranging, the theoretical velocity through the orifice, v, is therefore

$$v = \sqrt{2gH}$$

and is known as the Torricelli equation. The actual velocity v_{act}, however, is less than this theoretical prediction due to permanent and irreversible energy losses and can be found by introducing the coefficient of velocity C_v, which represents the ratio of the actual to theoretical jet velocity. That is

$$C_v = \frac{v_{act}}{v}$$

As liquid emerges from the orifice, the streamlines converge to form a *vena contracta* just beyond the orifice, at a distance of half the orifice diameter. The cross-sectional area of the *vena contracta* can be found by introducing a coefficient of contraction C_c, and is the ratio of actual to theoretical flow area. The coefficient of discharge C_d is therefore defined as the product of C_v and C_c where

$$\begin{aligned} Q_{act} &= C_v C_c a_o \sqrt{2gH} \\ &= C_d a_o \sqrt{2gH} \end{aligned}$$

The total flow through all the orifices is therefore

$$Q = nC_d a_o \sqrt{2gH}$$

where n is the number of orifices. Rearranging

$$n = \frac{Q}{C_d a_o \sqrt{2gH}}$$

$$= \frac{\dfrac{5000}{3600 \times 24}}{0.6 \times \dfrac{\pi \times 0.01^2}{4} \times \sqrt{2 \times g \times 0.2}}$$

$$= 620$$

A total of 620 orifices is required. Note that at the *vena contracta* the streamlines are no longer converging as the flow passes through the orifices, but are parallel. The internal pressure in the jet will have decreased from its upstream pressure to atmospheric which is imposed on the free jet. At this point, the velocity across the jet is essentially uniform and equal to the total head.

6.2 Coefficient of velocity

To determine the coefficient of velocity of a small circular sharp-edged orifice in the side of a vessel, the horizontal and vertical co-ordinates of the trajectory of the jet were measured for a head of 20 cm. The horizontal co-ordinate from the vena contracta *was found to be 86 cm whilst the vertical co-ordinate was 96 cm. Determine the coefficient of velocity.*

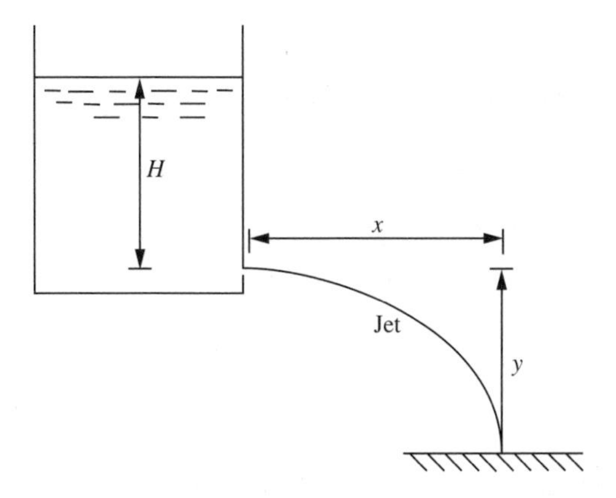

Solution

The coefficient of velocity for sharp-edged orifices located in the side of a tank can be readily determined from the trajectory of the issuing free jet, in which the horizontal distance travelled by the jet measured from the *vena contracta* is

$$x = v_{act} t$$

through a vertical distance

$$y = \frac{1}{2} g t^2$$

Eliminating time t, the actual velocity of the jet is

$$v_{act} = x\sqrt{\frac{g}{2y}}$$

The theoretical velocity is given by Torricelli's equation as

$$v = \sqrt{2gH}$$

The coefficient of velocity is therefore

$$\begin{aligned} C_v &= \frac{v_{act}}{v} \\ &= \frac{x\sqrt{\frac{g}{2y}}}{\sqrt{2gH}} \\ &= \frac{x}{2\sqrt{yH}} \\ &= \frac{0.86}{2 \times \sqrt{0.96 \times 0.2}} \\ &= 0.981 \end{aligned}$$

The coefficient of velocity for the orifice is found to be 0.981. Note that the relationship between velocity and head which was first developed by Torricelli in around 1645 is a simple treatment of the Bernoulli equation in which static head is converted to velocity head. Since the static head varies uniformly with depth in the vessel, it is possible to predict the trajectory of the free jet from sharp-edged orifices for other elevations on the side of the vessel.

The velocity of the discharging jet is assumed to be uniform at the *vena contracta* (see Problem 6.1, page 143). This is true for sharp-edged orifices but not for rounded orifices. Rounded orifices have the effect of slightly reducing the coefficient of discharge. Significantly, the coefficient of contraction is more marked with sharp-edged orifices. Precise measurements of the area of the *vena contracta*, however, are not always possible. Better estimates can usually be obtained from estimates of the coefficient of discharge and coefficient of velocity such that $C_c = C_d / C_v$.

6.3 Drainage from tank with uniform cross-section

A cylindrical tank of diameter 1 m mounted on its axis contains a liquid which drains through a 2 cm diameter hole in the base of the tank. If the tank originally contains 1000 litres, determine the time taken for total drainage. The discharge coefficient for the hole may be taken as 0.6.

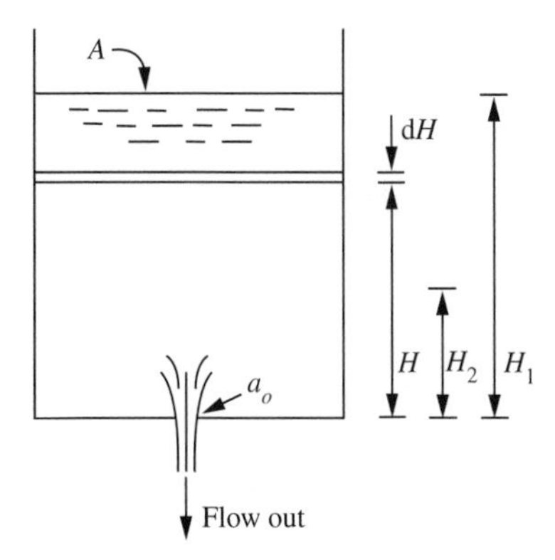

Solution

Consider a tank of uniform cross-section A_t, through an orifice of area a_o, located in the base of the tank. An unsteady state mass balance on the tank for no inflow of liquid relates the rate of flow from the tank through the orifice to the change of capacity of the tank

$$C_d a_o \sqrt{2gH} = -A \frac{dH}{dt}$$

Rearranging, the time taken for tank drainage from a head H_1 to H_2 can be found by integration

$$\int_0^t dt = \frac{-A}{C_d a_o \sqrt{2g}} \int_{H_1}^{H_2} H^{-\frac{1}{2}} dH$$

Completing the integration

$$t = \frac{-A}{C_d a_o} \sqrt{\frac{2}{g}} \left(H_2^{\frac{1}{2}} - H_1^{\frac{1}{2}} \right)$$

For total drainage, $H_2 = 0$. Therefore

$$t = \frac{A}{C_d a_o}\sqrt{\frac{2H_1}{g}}$$

The initial liquid level above the orifice, H_1, is obtained from the volume of the tank V_t, where d_t is the diameter of the tank

$$H_1 = \frac{4V_t}{\pi d_t^2}$$

$$= \frac{4 \times 1}{\pi \times 1^2}$$

$$= 1.27 \text{ m}$$

Therefore

$$t = \frac{\dfrac{\pi \times 1^2}{4}}{0.6 \times \dfrac{\pi \times 0.02^2}{4}} \times \sqrt{\frac{2 \times 1.27}{g}}$$

$$= 2120 \text{ s}$$

Total drainage is found to take 35 minutes and 20 seconds.

6.4 Tank drainage with hemispherical cross-section

A hemispherical tank of 4 m diameter contains a liquid and is emptied through a hole of diameter 5 cm. If the discharge coefficient for the hole is 0.6, determine the time required to drain the tank from an initial depth of 1.5 m.

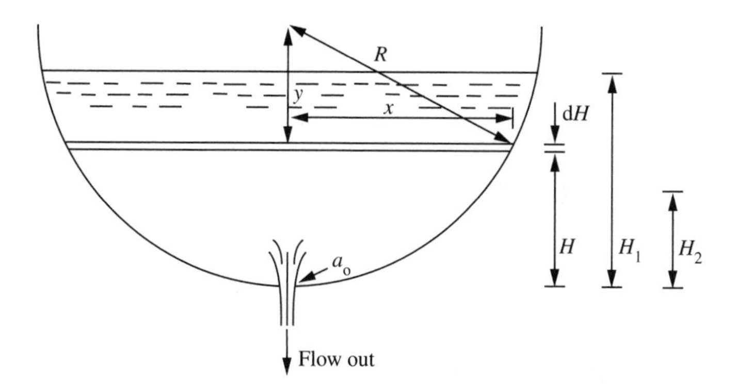

Solution

Applying an unsteady mass state balance on the tank, the rate of discharge from the hole is equal to the change in capacity in the tank

$$C_d a_o \sqrt{2gH} = -A\frac{dH}{dt}$$ (see Problem 6.3, page 147)

The depth of liquid in the tank does not change uniformly with cross-sectional area of the tank. The depth H is related to the surface area of the liquid in the tank where

$$A = \pi x^2$$

The distance x is related to depth H by Pythagoras

$$R^2 = x^2 + y^2$$

where

$$R = y + H$$

Therefore

$$x^2 = R^2 - (R - H)^2$$

$$= 2RH - H^2$$

The balance therefore becomes

$$\int_0^t \mathrm{d}t = \frac{-\pi}{C_d a_o \sqrt{2g}} \int_{H_1}^{H_2} \frac{(2RH - H^2)}{H^{\frac{1}{2}}} \mathrm{d}H$$

$$= \frac{-\pi}{C_d a_o \sqrt{2g}} \int_{H_1}^{H_2} (2RH^{\frac{1}{2}} - H^{\frac{3}{2}}) \mathrm{d}H$$

Therefore on integration

$$t = \frac{\pi}{C_d a_o \sqrt{2g}} \left(\frac{4}{3} R(H_1^{\frac{3}{2}} - H_2^{\frac{3}{2}}) - \frac{2}{5}(H_1^{\frac{5}{2}} - H_2^{\frac{5}{2}}) \right)$$

To empty the tank, $H_2 = 0$. Therefore

$$t = \frac{\pi}{C_d a_o \sqrt{2g}} \left(\frac{4}{3} RH_1^{\frac{3}{2}} - \frac{2}{5} H_1^{\frac{5}{2}} \right)$$

$$= \frac{\pi}{0.6 \times \frac{\pi \times 0.05^2}{4} \times \sqrt{2g}} \times \left(\frac{4}{3} \times 2 \times 1.5^{\frac{3}{2}} - \frac{2}{5} \times 1.5^{\frac{5}{2}} \right)$$

$$= 2286 \text{ s}$$

The time to drain the tank is found to be approximately 38 minutes. Note that if the vessel were initially full and is completely emptied, then

$$H_1 = R$$

$$H_2 = 0$$

to give

$$t = \frac{14\pi R^{\frac{5}{2}}}{15 C_d a \sqrt{2g}}$$

6.5 Tank drainage with cylindrical cross-section

Derive an expression for the time to discharge liquid from a depth H in a cylindrical tank of radius R and length L positioned on its side through an orifice of area a_o located on the underside of the tank.

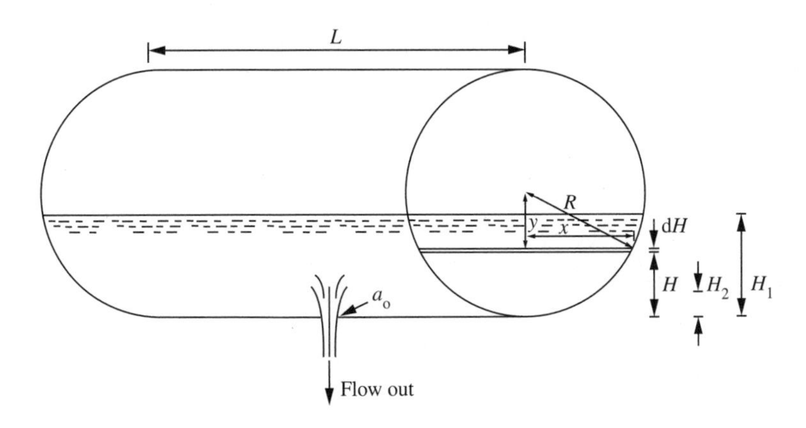

Solution

The time taken for the tank to drain can be determined from

$$C_d a_o \sqrt{2gH} = -A\frac{dH}{dt}$$ (see Problem 6.3, page 147)

The area of the liquid surface varies with depth where

$$A = 2xL$$

The distance x is be related to liquid depth by Pythagoras

$$R^2 = x^2 + y^2$$

where

$$R = y + H$$

Therefore

$$x = \sqrt{2RH - H^2}$$

The balance equation is therefore

$$C_d a_o \sqrt{2gH} = -2\sqrt{2RH - H^2}\, L \frac{\mathrm{d}H}{\mathrm{d}t}$$

Rearranging, the total time for discharge is found by integrating

$$\int_0^t \mathrm{d}t = \frac{-2L}{C_d a_o \sqrt{2g}} \int_{H_1}^{H_2} (2R - H)^{\frac{1}{2}}\, \mathrm{d}H$$

to give

$$t = \frac{4L}{3C_d a_o \sqrt{2g}} \left((2R - H_2)^{\frac{3}{2}} - (2R - H_1)^{\frac{3}{2}} \right)$$

Note that if the vessel was half full and was completely emptied through the orifice then

$$H_1 = R$$

$$H_2 = 0$$

then the time for drainage would be

$$t = \frac{4LR^{\frac{3}{2}}}{3C_d a_o \sqrt{2g}}$$

6.6 Drainage between two reservoirs

A reservoir beneath a small forced convection water cooling tower is 4 m long and 2 m wide. Before recirculation, the water flows into a smaller adjacent reservoir of dimensions 2 m by 2 m by way of a submerged opening with a flow area of 100 cm² and discharge coefficient of 0.62. During normal operation the level in the larger reservoir is 30 cm above that in the smaller reservoir. Determine the rate of flow between the reservoirs during normal operation and if the recirculation is halted, determine the time taken to reduce the difference in levels to 10 cm.

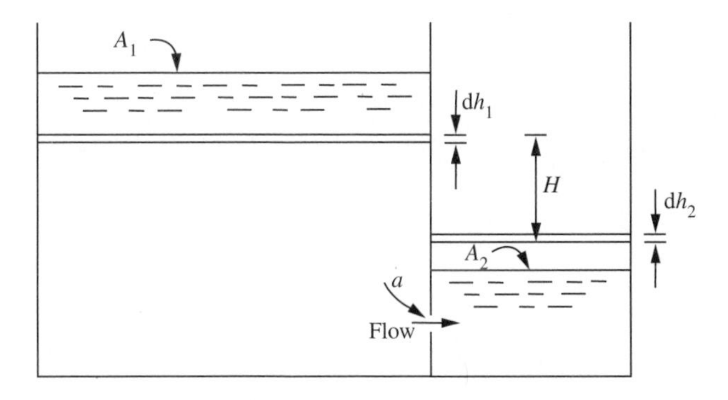

Solution

For steady operation with a continuous flow between reservoirs the rate of flow through the submerged opening is given by

$$Q = C_d a \sqrt{2gH}$$

$$= 0.62 \times 0.01 \times \sqrt{2g \times 0.3}$$

$$= 0.015 \text{ m}^3\text{s}^{-1}$$

Thus, the flow under steady conditions is found to be 0.015 m^{3}s^{-1}. When the recirculation is halted the drainage from the larger tank to the smaller tank proceeds and the difference in levels H changes by increments of dH while in the larger and smaller tanks by dh_1 and dh_2, respectively

$$H - \mathrm{d}H = H - \mathrm{d}h_1 - \mathrm{d}h_2$$

The incremental loss in capacity of the larger tank is equal to the incremental gain in capacity in the smaller tank. That is

$$A_1 \mathrm{d}h_1 = A_2 \mathrm{d}h_2$$

The incremental difference in levels is therefore

$$\mathrm{d}H = \mathrm{d}h_1 \left(1 + \frac{A_1}{A_2} \right)$$

The change in capacity of the larger tank is given by the unsteady state equation (see Problem 6.3, page 147)

$$C_d a \sqrt{2gH} = -A_1 \frac{\mathrm{d}h_1}{\mathrm{d}t}$$

Substituting for $\mathrm{d}h_1$

$$C_d a \sqrt{2gH} = \frac{-A_1}{\left(1 + \frac{A_1}{A_2} \right)} \frac{\mathrm{d}H}{\mathrm{d}t}$$

Rearranging

$$\int_0^t \mathrm{d}t = \frac{-A_1}{C_d a \sqrt{2g} \left(1 + \frac{A_1}{A_2} \right)} \int_{H_1}^{H_2} H^{-\frac{1}{2}} \mathrm{d}H$$

Integrating gives

$$t = \frac{2A_1 (H_1^{\frac{1}{2}} - H_2^{\frac{1}{2}})}{C_d a \sqrt{2g} \left(1 + \frac{A_1}{A_2} \right)}$$

$$= \frac{2 \times 4 \times 2 \times (0.3^{\frac{1}{2}} - 0.1^{\frac{1}{2}})}{0.62 \times 0.01 \times \sqrt{2g} \times \left(1 + \frac{4 \times 2}{2 \times 2} \right)}$$

$$= 45 \text{ s}$$

The time taken is found to be 45 seconds.

6.7 Tank inflow with simultaneous outflow

Derive an expression for the time taken for the liquid in a tank to reach a new level if the liquid drains from an orifice while there is a constant flow of liquid into the tank.

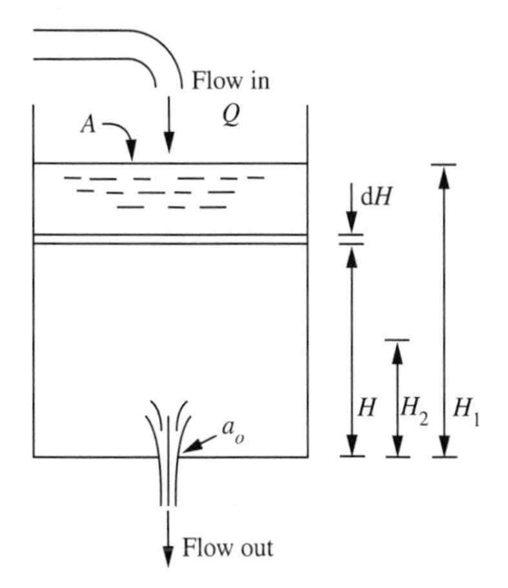

Solution

The unsteady state mass balance on the tank is

$$Q = C_d a_o \sqrt{2gH} + A\frac{dH}{dt}$$ (see Problem 6.3, page 147)

where Q is the rate of flow into the tank and dH/dt is the rate of change in liquid level. For convenience, k is used to group the constants

$$k = C_d a_o \sqrt{2g}$$

Therefore

$$Q - k\sqrt{H} = A\frac{dH}{dt}$$

To determine the head of liquid in the tank at any given time, it is therefore necessary to integrate between the limits H_1 and H_2

$$\int_0^t dt = A\int_{H_1}^{H_2} \frac{dH}{Q - kH^{\frac{1}{2}}}$$

To simplify this integration the substitution is used

$$u = Q - kH^{\frac{1}{2}}$$

Rearranging

$$H = \left(\frac{Q - u}{k}\right)^2$$

and differentiating with respect to u

$$\mathrm{d}H = \frac{-2(Q - u)\mathrm{d}u}{k^2}$$

The integration becomes

$$\int_0^t \mathrm{d}t = \frac{-2A}{k^2}\int \frac{Q - u}{u}\mathrm{d}u$$

$$= \frac{-2A}{k^2}\left(Q\int \frac{\mathrm{d}u}{u} - \int \mathrm{d}u\right)$$

Completing the integration gives

$$t = \frac{-2A}{k^2}\left(Q \log_e\left[\frac{Q - kH_2^{\frac{1}{2}}}{Q - kH_1^{\frac{1}{2}}}\right] + k(H_2^{\frac{1}{2}} - H_1^{\frac{1}{2}})\right)$$

or in full

$$t = \frac{-2A}{(C_d a_o \sqrt{2g})^2}\left(Q \log_e\left[\frac{Q - C_d a_o \sqrt{2g\, H_2}}{Q - C_d a_o \sqrt{2g\, H_1}}\right] + C_d a_o \sqrt{2g}(H_2^{\frac{1}{2}} - H_1^{\frac{1}{2}})\right)$$

6.8 Instantaneous tank discharge

A concrete tank is 15 m long by 10 m wide and its sides are vertical. Water enters the tank at a rate of 200 litres per second and is discharged from a sluice, the centre line of which is 50 cm above the bottom of the tank. When the depth of water in the tank is 2 m, the instantaneous rate of discharge is observed to be 400 litres per second. Determine the time for the level in the tank to fall 1 m.

Solution

For a tank of uniform cross-section which receives a steady flow of liquid at a rate Q and is allowed to drain freely through an opening or sluice, the time taken to alter the level can be shown (see Problem 6.7, page 155) to be

$$t = \frac{-2A}{k^2}\left(Q \log_e \left[\frac{Q - kH_2^{\frac{1}{2}}}{Q - kH_1^{\frac{1}{2}}} \right] + k(H_2^{\frac{1}{2}} - H_1^{\frac{1}{2}}) \right)$$

The discharge coefficient and the area of the sluice are not provided. However, the rate of flow through the sluice Q_s is related to head by

$$Q_s = kH^{\frac{1}{2}}$$

or

$$k = Q_s H^{-\frac{1}{2}}$$

When the level was 2.0 m the flow was 0.4 m^3s^{-1}. That is a depth of liquid above the centre line of the sluice, H_1, of 1.5 m. Thus

$$k = 0.4 \times 1.5^{-\frac{1}{2}}$$

$$= 0.326 \text{ m}^{\frac{5}{2}}\text{s}^{-1}$$

The final depth, H_2, is 0.5 m above the centre line of the sluice. The time taken for the level to fall is therefore

$$t = \frac{-2A}{k^2}\left(Q \log_e\left[\frac{Q - kH_2^{\frac{1}{2}}}{Q - kH_1^{\frac{1}{2}}}\right] + k(H_2^{\frac{1}{2}} - H_1^{\frac{1}{2}})\right)$$

$$= \frac{-2 \times 15 \times 10}{0.326^2} \times \left(0.2 \times \log_e\left[\frac{0.2 - 0.326 \times 0.5^{\frac{1}{2}}}{0.2 - 0.326 \times 1.5^{\frac{1}{2}}}\right] + 0.326 \times (0.5^{\frac{1}{2}} - 1.5^{\frac{1}{2}})\right)$$

$$= 1536 \text{ s}$$

The time is found to be approximately 25½ minutes.

Note that although the area of the sluice and discharge coefficient are not provided, the constant k is used to substitute

$$k = C_d a \sqrt{2g}$$

Assuming a reasonable discharge coefficient for the sluice of 0.62, the area of the sluice a can be found by rearranging

$$a = \frac{k}{C_d \sqrt{2g}}$$

$$= \frac{0.326}{0.62 \times \sqrt{2g}}$$

$$= 0.119 \text{ m}^2$$

This would correspond to a sluice of square cross-section with sides measuring 34 cm. Care should be taken, however, when considering discharges through large orifices where the head producing flow may be substantially less at the top than at the bottom.

6.9 Instantaneous tank inflow with outflow

A tank of uniform cross-section is provided with a circular orifice 50 mm in diameter in the bottom. Water flows into the tank at a uniform rate and is discharged through the orifice. It is noted that it takes 90 seconds for the head in the tank to rise from 60 cm to 70 cm and 120 seconds for it to rise from 120 cm to 125 cm. Determine the rate of inflow and the cross-sectional area of the tank assuming a discharge coefficient of 0.62 for the orifice.

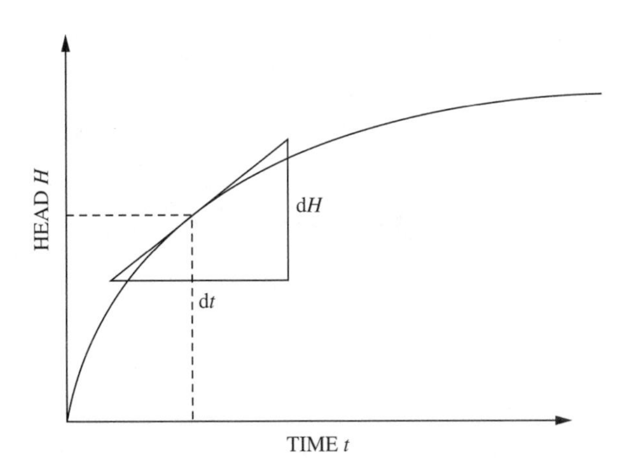

Solution

An unsteady state mass balance on the tank which receives a flow of water at a constant rate Q is

$$Q = C_d a_o \sqrt{2gH} + A\frac{dH}{dt}$$ (see Problem 6.7, page 155)

Rearranging

$$\frac{dH}{dt} = \frac{Q}{A} - \frac{C_d a_o \sqrt{2gH}}{A}$$

$$= \frac{Q}{A} - \frac{kH^{\frac{1}{2}}}{A}$$

where

$$k = C_d a_o \sqrt{2g}$$

$$= 0.62 \times \frac{\pi \times 0.05^2}{4} \times \sqrt{2g}$$

$$= 5.39 \times 10^{-3}\ \text{m}^{\frac{5}{2}}\text{s}^{-1}$$

For an average head of 0.65 m, the change in level is 0.1 m over 110 seconds. That is

$$\frac{0.1}{90} = \frac{dH}{dt}$$

$$= \frac{Q}{A} - \frac{5.39 \times 10^{-3} \times 0.65^{\frac{1}{2}}}{A}$$

For an average head of 1.225 m, the change in level is 0.05 m taking 120 seconds. That is

$$\frac{0.05}{120} = \frac{dH}{dt}$$

$$= \frac{Q}{A} - \frac{5.39 \times 10^{-3} \times 1.225^{\frac{1}{2}}}{A}$$

Solving the simultaneous equations

$$A = 2.34\ \text{m}^2$$

$$Q = 6.94 \times 10^{-3}\ \text{m}^3\text{s}^{-1}$$

The area of the tank and flowrate are found to be 2.34 m^2 and 6.94 litres per second, respectively.

6.10 Tank discharge through a horizontal pipe with laminar flow

A viscous Newtonian liquid of density 1100 kgm^{-3} and viscosity 0.08 Nsm^{-2} is fed to a process from a vessel of diameter 1.2 m through a 3 m length of horizontal pipe with an inside diameter of 25 mm attached near to the base of the vessel. If the initial level of liquid in the vessel is 1.5 m above the pipe and the flow through the pipe is laminar, determine the time to feed 0.75 m^3 of the liquid to the process.

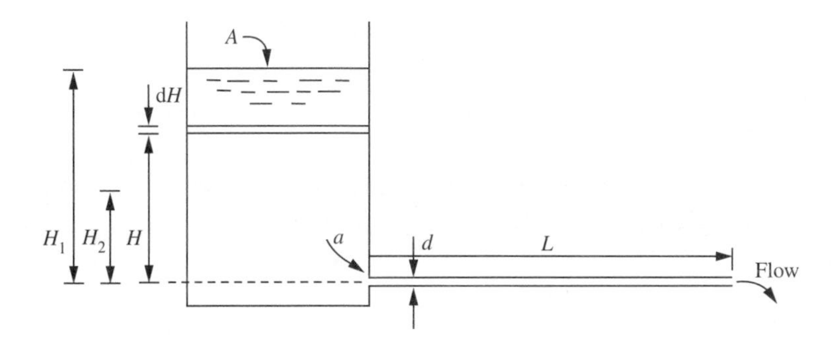

Solution

Neglecting entrance and exit losses, an unsteady mass state balance for the vessel is

$$av = -A\frac{dH}{dt}$$

If the head in the vessel is used to overcome the viscous resistance of the pipe, the average velocity of the flowing liquid is therefore given by

$$v = \frac{\Delta p d^2}{32\mu L}$$

(see Problem 3.7, page 73)

$$= \frac{\rho g H d^2}{32\mu L}$$

Therefore

$$\frac{a\rho gHd^2}{32\mu L} = -A\frac{\mathrm{d}H}{\mathrm{d}t}$$

The total time taken for the level to fall from a head H_1 to H_2 is therefore found from

$$\int_0^t \mathrm{d}t = \frac{-32\mu LA}{a\rho gd^2}\int_{H_1}^{H_2}\frac{\mathrm{d}H}{H}$$

On integration this becomes

$$t = \frac{-32\mu LA}{a\rho gd^2}\log_e\left[\frac{H_2}{H_1}\right]$$

where the final level in the vessel, H_2, is found from the total volume discharged to the process

$$H_2 = H_1 - \frac{4V}{\pi d_t^2}$$

$$= 1.5 - \frac{4\times 0.75}{\pi\times 1.2^2}$$

$$= 0.84 \text{ m}$$

The time taken is therefore

$$t = \frac{-32\times 0.08\times 3\times\dfrac{\pi\times 1.2^2}{4}}{\dfrac{\pi\times 0.025^2}{4}\times 1100\times g\times 0.025^2}\times\log_e\left[\frac{0.84}{1.5}\right]$$

$$= 1521 \text{ s}$$

The time is found to be approximately 25½ minutes.

Further problems

(1) State the assumptions upon which simple tank drainage problems are based.

(2) A cylindrical vessel mounted vertically on its axis has a cross-sectional area of 1.2 m^2 and contains water which is allowed to discharge freely through an orifice, with a cross-sectional area of 100 cm^2, positioned in the side of the tank 4 m below the surface. If the tank is open to atmosphere, determine the time taken for the level to fall by 2 m. The discharge coefficient may be taken as 0.6.

Answer: 53 s

(3) A rectangular orifice in the side of a tank is 1 m broad and 50 cm deep. The level of water in the tank is 50 cm above the top edge of the orifice. Determine the flow through the orifice if the coefficient of discharge is 0.62. Note that the velocity through the orifice may not be regarded as constant as the variation in head at different depths of the orifice will be considerable.

Answer: 1.18 m^3s^{-1}

(4) A rectangular orifice in the side of a large water tank has a breadth of 1 m and depth of 0.5 m. The water on one side of the orifice is at a depth of 1 m above the top edge; the water level on the other side of the orifice is 25 cm below the top edge. Determine the rate of discharge if the discharge coefficient may be taken as 0.62.

Answer: 1.49 m^3s^{-1}

(5) A spherical storage vessel 3 m in diameter contains a process liquid at half capacity. Determine the time taken to drain the vessel to a depth of 1 m through an orifice with a diameter of 25 mm at the bottom of the vessel. The discharge coefficient of the orifice is 0.6.

Answer: 2341 s

(6) Derive an expression for the time to drain an open hemispherical vessel of radius R through an orifice of area a_o, if the vessel has an inflow of liquid of constant rate Q.

(7) Show that the time taken to drain a spherical vessel of radius R completely from full through an orifice of area a located at the bottom of the vessel is

$$t = \frac{14\pi R^{\frac{5}{2}}}{15C_d a\sqrt{2g}}$$

where C_d is the coefficient of discharge and g the acceleration due to gravity.

(8) A cylindrical vessel of diameter 3 m vertically mounted on its axis contains a liquid at an initial depth of 4 m and is allowed to discharge to a similar nearby vessel of diameter 4 m by way of a pipe with an inside diameter of 100 mm. Determine the final depth in both vessels and the time to reach this condition. Ignore any losses due to friction.

Answer: 1.44 m, 502 s

(9) A rectangular tank 6 m long by 2 m wide is divided in two parts by a partition so the area of one part is twice the area of the other. The partition contains a sluice of area 100 cm^2. If the level of liquid in the smaller division is 2 m above that of the larger, determine the time to reduce the difference in level to 50 cm. Assume a discharge coefficient of 0.6.

Answer: 142 s

(10) Two identical open tanks of cross-sectional area 4 m^3 are connected by a straight length of pipe of length 10 m and internal diameter 5 cm. A viscous oil with a viscosity of 0.1 Nsm^{-2} and density 900 kgm^{-3} is initially at a depth of 2 m in one of the tanks while the other tank is empty. If a valve in the connecting pipe is fully opened, determine the time for the difference in level to fall to 5 mm. Neglect pipe entrance and exit losses and assume the connecting pipe is initially full of oil.

Answer: 5100 s

(11) In an experiment to determine the coefficients of contraction, velocity and discharge for a circular orifice of 9 mm diameter, water was discharged through the orifice mounted vertically in the side of a tank. A constant head of water of 1.2 m was maintained above the centre line of the orifice. The issuing jet was found to strike a target plate a horizontal distance of 850 mm from the *vena contracta* a vertical distance of 155 mm below the centre line of the orifice. Determine the values of the coefficients for the orifice if the measured discharge from the orifice was 91 litres in 470 seconds.

Answer: 0.636, 0.985, 0.627

(12) Show that the time taken for liquid of density ρ and viscosity μ to drain with laminar flow from tank of diameter d_t and depth H through a vertical pipe of inside diameter d and length L, attached to the underside of the tank is given by

$$t = \frac{32\mu L d_t^2}{\rho g d^4} \log_e\left(\frac{H + L}{L}\right)$$

(13) Show, where the area of the tank is not appreciably greater than the area of the orifice, that the theoretical velocity of a jet of liquid v, for a constant head H, flowing from an orifice in the side of a tank, can be given by

$$v = \sqrt{\frac{2gH}{1 - \left(\frac{a}{A}\right)^2}}$$

where a is the area of the orifice and A is the cross-sectional area of the tank.

(14) The bottom of a process vessel has a conical section and contains a liquid which is required to be drained through an opening with a diameter of 25 mm. If the liquid in the section is initially at a depth of 1.5 m above the opening corresponding to a diameter of the section of 1.73 m, determine the time to drain the section completely. State any assumptions used.

Answer: 883 s

(15) A process vessel with a uniform area of 1.2 m^2 receives a liquid at a steady rate of 0.04 m^3s^{-1} and is simultaneously discharged through an opening at the bottom of the vessel. When the depth of liquid was 0.5 m the instantaneous rate of discharge was noted as being 0.03 m^3s^{-1}. Determine the time taken for the level to rise by 20 m.

Answer: 70 s

(16) Water flows at a steady rate of 360 m^3h^{-1} into a vertical-sided tank of area 10 m^2. The water discharges continuously from a sluice of area 0.0564 m^2. If the initial level in the tank is 2.5 m above the sluice, determine the final depth after 5 minutes assuming a discharge coefficient of 0.6 for the sluice.

Answer: 0.6 m

Open channels, notches and weirs

7

Introduction

Unlike the single-phase flow of fluids through pipes, the flow of liquids through open channels or flumes is characterized by a free surface normally at, or near, atmospheric pressure. Rivers and artificial canals are examples of open channels. Open channel flow also applies, however, to the flow of liquids through pipes which are not run full, as in the case of sewerage pipes. Weirs are vertical obstructions which lie across open channels, whilst notches are openings, normally rectangular or triangular, cut in such weirs. In practice, the liquid in the channel builds up behind the weir until flow over the weir occurs with an equilibrium head of liquid being a measure of flowrate.

There are numerous applications of open channel flow, notches and weirs. Although primarily encountered in the process effluent industry, there are many other applications in the chemical, power and oil industries. The perforated trays of distillation columns, for example, involve the open flow of liquid through which vapour is bubbled up from the tray beneath, with the depth of liquid maintained by a weir over which the liquid discharges into the downcomer.

Many sophisticated mathematical methods and procedures have been developed and applied to various types of open channel flow. Traditional methods are still widely used, however, based on the pioneering work of the French engineer Antoine Chézy in 1775 and the experimental work of the Irish engineer Robert Manning in 1891. In its simplest form, the channel is assumed to have a gentle slope in which the rate of flow under the influence of gravity is balanced by frictional losses and such that the streamlines in the liquid run parallel to the floor (or bed) of the channel. Under conditions of uniform flow, there is therefore no change in velocity in the flowing liquid (no acceleration) and the depth is constant. Unlike single-phase pipe flow, the pressure conditions within the liquid are determined by hydrostatic principles and constant atmospheric conditions at the free surface.

The geometry of channels varies from wide and shallow to narrow and deep with side walls that may be vertical, inclined at an angle or curved. As open channels have a free surface, open channel flow therefore differs from pipe flow in that the cross-section of flow is not constrained. For increased rate of flow there is an increased depth. This affects not only the flow area but also wetted surface (both bed and wall) of the channel. An important parameter used to characterize the geometry of channels is the mean hydraulic depth defined as the ratio of flow area to wetted perimeter.

For problems involving open channels, determining the rate of flow for a given depth of liquid is straightforward. Arriving at the depth for a given flowrate, however, usually requires trial and error approaches since the depth influences both the flow area and the wetted surface. The common method is to guess a value for the depth and calculate the corresponding flowrate, repeating the procedure a sufficient number of times to arrive at the answer.

It should be noted that many of the equations associated with open channel flow are empirically based. Many of the available data have been obtained using water as the liquid medium. Empirical relationships used to predict the flow for other liquids should therefore be applied with caution.

7.1 Chézy formula for open channel flow

Derive the Chézy formula for the uniform flow of a liquid along an open channel inclined with a slope i.

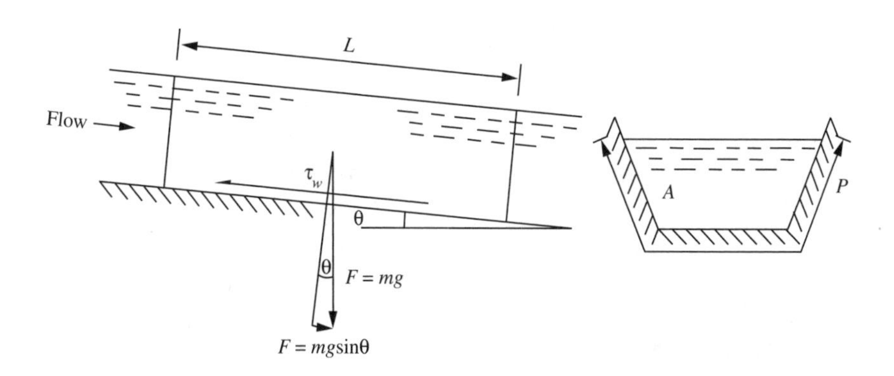

Solution

Consider a liquid of density ρ flowing in an open channel of uniform cross-sectional area A, inclined at a small angle θ. Under steady uniform flow, the gravitational force perpendicular to the base of the channel is balanced by the viscous resistance forces acting in the opposite direction. The mass of the liquid in the channel can be expressed in terms of liquid density and volume

$$F = mg \sin \theta$$

$$= \rho A L g \sin \theta$$

The resistance forces which retard flow are obtained by considering the mean wall shear stress τ_w of the liquid in contact with the wetted channel wall.

$$F = \tau_w PL$$

For equilibrium

$$\rho A L g \sin \theta = \tau_w PL$$

If the mean wall shear stress can be related to the kinetic energy per unit volume

$$\tau_w = \frac{f}{2} \rho v^2$$

where f is a form of friction coefficient. Noting also that for small angles, sin θ is approximately equal to tan θ or the slope of the channel i, and defining the ratio of flow area A to wetted perimeter P as the mean hydraulic depth

$$m = \frac{A}{P}$$

the average velocity of the liquid flowing in the channel can therefore be expressed as

$$v = \sqrt{\frac{2gmi}{f}}$$

That is, for a given friction factor and mean hydraulic depth the velocity is proportional to the square root of the slope i. Defining also

$$C = \sqrt{\frac{2g}{f}}$$

The flow through the channel may therefore be given by

$$Q = CA\sqrt{mi}$$

This is known as the Chézy formula where C is the Chézy coefficient. Experimental values and several correlations are available, the most common of which is the Manning formula given by

$$C = \frac{m^{\frac{1}{6}}}{n}$$

where n is a dimensional roughness factor, the magnitude of which depends on the type of surface. Examples of typical roughness factors are given on page 292. Note that as the roughness of the channel increases, the value of n also increases, reducing the value of the Chézy coefficient and therefore flow.

7.2 Flow in a rectangular open channel

Water for cooling is delivered to a cooling tower along a rectangular concrete channel 1 m wide. Determine the rate of delivery for uniform flow at a depth of 30 cm. The slope of the channel is 1:1000 and the roughness factor for the concrete is 0.014 $m^{-1/3}s$.

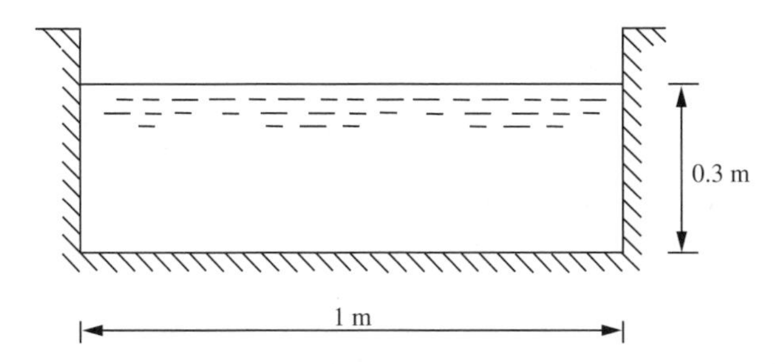

Solution

The delivery of water through the open channel is given by the Chézy formula

$$Q = CA\sqrt{mi}$$ (see Problem 7.1, page 169)

where i is the slope, A is the flow area and C is the Chézy coefficient given by the Manning formula (see Problem 7.1, page 169). The discharge is thus

$$Q = \frac{m^{\frac{1}{6}}}{n} A\sqrt{mi}$$

$$= \frac{A}{n} m^{\frac{2}{3}} \sqrt{i}$$

For the dimensions of the channel, the flow area is

$$A = 1 \times 0.3$$

$$= 0.3 \text{ m}^2$$

and wetted perimeter is

$$P = 2 \times 0.3 + 1$$

$$= 1.6 \text{ m}$$

The mean hydraulic depth is therefore

$$m = \frac{A}{P}$$

$$= \frac{0.3}{1.6}$$

$$= 0.1875 \text{ m}$$

Thus, the rate of delivery is

$$Q = \frac{0.3}{0.014} \times 0.1875^{\frac{2}{3}} \times \sqrt{0.001}$$

$$= 0.222 \text{ m}^3\text{s}^{-1}$$

The delivery of water through the channel is found to be 0.222 m^3s^{-1}.

7.3 Depth of flow in a rectangular channel

A concrete-lined rectangular channel 12 m wide has a slope of 1 in 10,000. Determine the depth of water flowing in it if the volumetric flowrate is 60 m^3s^{-1}. Obtain the Chézy coefficient from the Manning formula taking the roughness factor as 0.015 $m^{-1/3}s$.

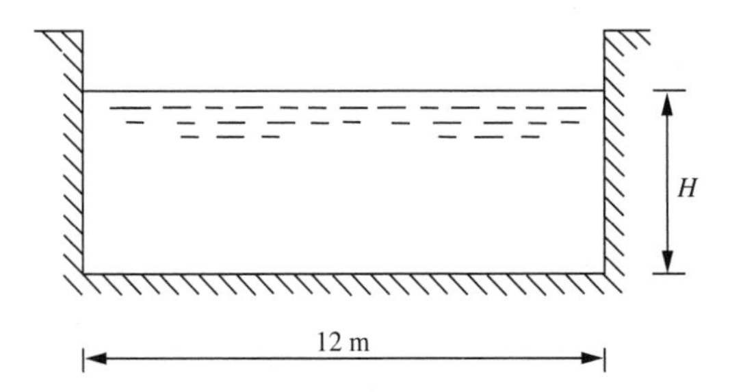

Solution

The flow through the channel is given by the Chézy formula

$$Q = CA\sqrt{mi}$$ (see Problem 7.1, page 169)

where i is the slope and C is the Chézy coefficient given by the Manning formula

$$C = \frac{m^{\frac{1}{6}}}{n}$$

where n is the roughness factor. The flow area of the channel in terms of depth is

$$A = 12H$$

and the wetted perimeter

$$P = 12 + 2H$$

The mean hydraulic depth (MHD) is therefore

$$m = \frac{A}{P}$$

$$= \frac{12H}{12 + 2H}$$

$$= \frac{6H}{6 + H}$$

An analytical solution for depth H is not possible. It is therefore necessary to use a graphical or trial and error approach illustrated below.

Head H, m	Area A, m^2	Perimeter P, m	MHD m, m	Flowrate Q, m^{3}s^{-1}
4.0	48	20	2.4	57.36
4.1	49.2	20.2	2.44	59.38
4.2	50.4	20.4	2.47	61.39
4.13	49.56	20.26	2.45	59.98

The depth is found to be approximately 4.13 m.

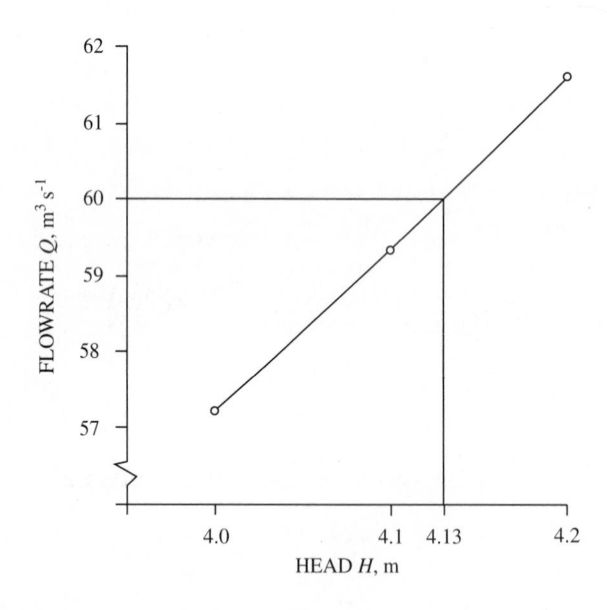

7.4 Economical depth of flow in rectangular channels

A rectangular channel is to be designed for conveying 300 m³ of water per minute. Determine the minimum cross-sectional area of the channel if the slope is 1 in 1600 and it can be assumed that

$$v = 70\sqrt{mi}$$

where v is the velocity of water in the channel, m is the mean hydraulic depth and i is the inclination of the channel.

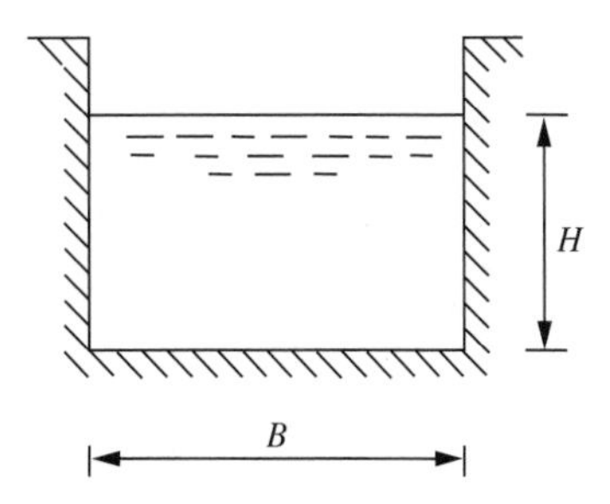

Solution

For a constant flow through the channel the maximum flow occurs when the wetted perimeter is a minimum. The wetted perimeter is

$$P = B + 2H$$

and flow area

$$A = HB$$

In terms of depth, the wetted perimeter is therefore

$$P = \frac{A}{H} + 2H$$

Differentiating the wetted perimeter with respect to depth and equating to zero

$$\frac{\mathrm{d}P}{\mathrm{d}H} = -\frac{A}{H^2} + 2$$

$$= 0$$

Therefore

$$A = 2H^2$$

Thus

$$B = 2H$$

That is, for the minimum cross-sectional area, the breadth of the channel is twice the channel depth. The mean hydraulic depth is therefore

$$m = \frac{A}{P}$$

$$= \frac{2H^2}{4H}$$

$$= \frac{H}{2}$$

From the equation provided for velocity, the flow through the channel is

$$Q = 70A\sqrt{mi}$$

In terms of depth H, the flow can therefore be expressed as

$$Q = 140H^{\frac{3}{2}}\sqrt{\frac{i}{2}}$$

Rearranging in terms of channel depth

$$H = \left(\frac{Q}{140}\sqrt{\frac{2}{i}}\right)^{\frac{2}{3}}$$

$$= \left(\frac{\left(\frac{300}{60}\right)}{140} \times \sqrt{2 \times 1600}\right)^{\frac{2}{3}}$$

$$= 1.6 \text{ m}$$

The flow area is therefore

$$A = 2H^2$$

$$= 2 \times 1.6^2$$

$$= 5.12 \text{ m}^2$$

The minimum flow area is found to be 5.12 m^2.

7.5 Circular channel flow

Determine the depth of flow of water in a sewer pipe of diameter 0.9144 m and inclination 1:200 when the discharge is 940 m^3h^{-1}. The Chézy coefficient may be taken as 100 $m^{1/2}s^{-1}$.

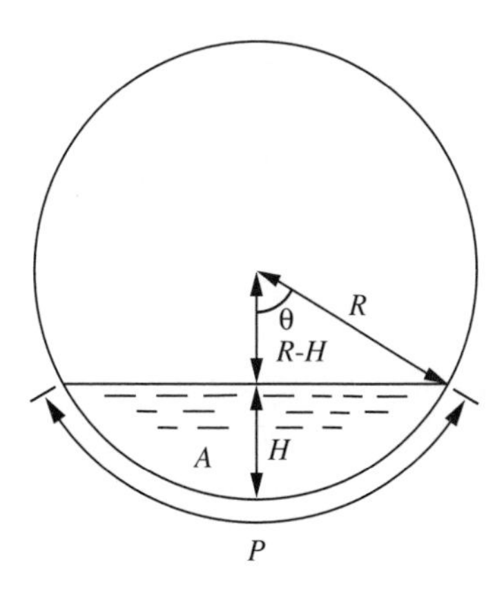

Solution

Let r be the radius, θ the half angle subtended at the centre by the water level and d the depth. From geometry

$$\cos\theta = \frac{R - H}{R}$$

from which the angle θ may be obtained in radians. The area of wetted cross-section is given by

$$A = \frac{2\theta R^2}{R} - R^2 \sin\theta\cos\theta$$

$$= R^2\left(\theta - \frac{\sin 2\theta}{2}\right)$$

and wetted perimeter is

$$P = 2R\theta$$

for which the mean hydraulic depth is

$$m = \frac{A}{P}$$

The flow through the channel is given by the Chézy formula

$$Q = CA\sqrt{mi} \qquad \text{(see Problem 7.1, page 169)}$$

where C is the Chézy coefficient and i is the channel slope. A graphical or trial and error approach is used for guessed values of depth:

H, m	$\cos\theta$	θ, rad	A, m^2	P, m	m, m	Q, m^3h^{-1}
0.1	0.781	0.674	0.039	0.616	0.063	248.7
0.15	0.671	0.834	0.070	0.763	0.092	543.2
0.2	0.562	0.973	0.106	0.890	0.119	932.2
0.25	0.453	1.101	0.146	1.006	0.145	1412.9
0.201	0.560	0.975	0.107	0.891	0.120	940.0

The depth in the channel is found to be 0.201 m.

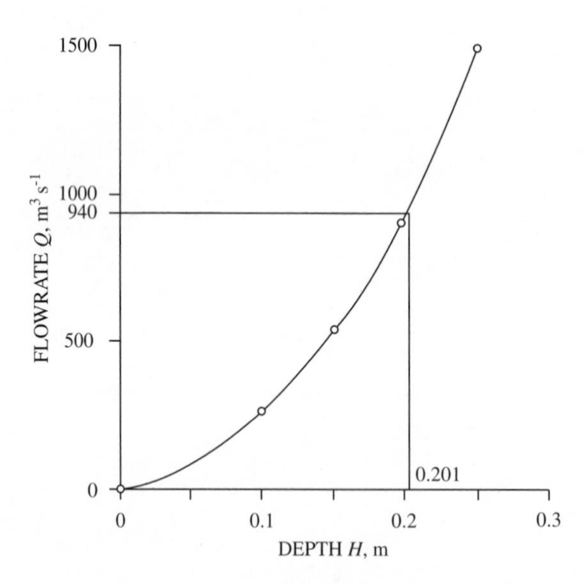

7.6 Maximum flow in circular channels

Determine the theoretical depth of liquid in a closed circular channel of radius R for maximum velocity.

Solution

The velocity of liquid in a circular channel depends on the depth. As the velocity is proportional to the mean hydraulic depth, its maximum value may be obtained by

$$P\frac{dA}{d\theta} - A\frac{dP}{d\theta} = 0$$

where from geometry, the wetted area of flow is

$$A = R^2\theta - R^2\frac{\sin 2\theta}{2}$$

(see Problem 7.5, page 177)

$$= R^2\left(\theta - \frac{\sin 2\theta}{2}\right)$$

and the wetted perimeter is

$$P = 2R\theta$$

Differentiating A/P and equating to zero

$$2R^3\theta(1 - \cos 2\theta) = 2R^3\left(\theta - \frac{\sin 2\theta}{2}\right)$$

Therefore

$$2\theta = \tan 2\theta$$

The solution to which is

$$2\theta = 257.5\,°$$

The depth for maximum velocity is therefore

$$H = R - R\cos\frac{257.5\,°}{2}$$

$$= R(1 + 0.62)$$

$$= 1.62R$$

The depth for maximum velocity is 0.81 times the channel diameter.

7.7 Weirs and rectangular notches

The weir above a downcomer in the rectifying section of a distillation column has a breadth of 1.2 m over which liquid flows with a head of 18 mm. Determine the rate of flow if the coefficient of discharge for the weir is 0.56.

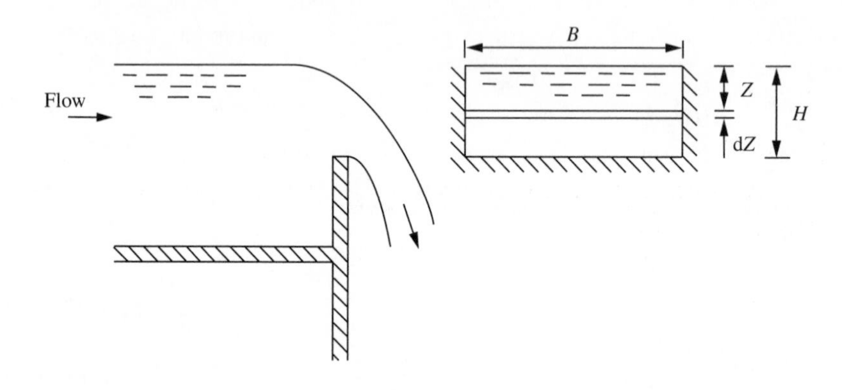

Solution

To determine the flowrate of liquid over weirs (and notches) consider the flow of liquid along a horizontal plane at depth z below the free surface. It is assumed that the upstream velocity, sometimes known as the velocity of approach, is negligible in comparison to the flow over the weir. At a point just beyond the weir, the velocity of the liquid will have increased to some value, v, in which the static head is converted to velocity head

$$z = \frac{v^2}{2g}$$

Therefore

$$v = \sqrt{2gz}$$

Thus, the velocity is proportional to the square root of the head. To determine the rate of flow, consider an elemental depth dz of the flow over the weir between depths z and $z + dz$. Since the elemental flow area is Bdz, the theoretical elemental flowrate dQ is therefore

$$dQ = \sqrt{2gz}\,Bdz$$

The total flowrate of liquid Q is found by integrating over the depth of the weir or rectangular notch H. That is

$$\int_0^Q \mathrm{d}Q = B\sqrt{2g}\int_0^H z^{\frac{1}{2}}\mathrm{d}z$$

to give

$$Q = \frac{2}{3}B\sqrt{2g}\,H^{\frac{3}{2}}$$

In practice, there are permanent energy losses due to edge effects giving rise to a flowrate which is lower than the theoretical. Introducing a discharge coefficient C_d, the actual flowrate is

$$Q = \frac{2}{3}C_d B\sqrt{2g}\,H^{\frac{3}{2}}$$

In this case

$$Q = \frac{2}{3} \times 0.56 \times 1.2 \times \sqrt{2g} \times 0.018^{\frac{3}{2}}$$

$$= 4.79 \times 10^{-3}\ \mathrm{m^3s^{-1}}$$

This corresponds to a rate of flow of liquid over the weir of 17.25 $\mathrm{m^3h^{-1}}$.

Note that the discharge coefficient takes into account the contraction of the overflowing jet as well as the effects of viscosity and surface tension (see Problem 4.2, page 102). Little is known of the separate influence of these two fluid parameters except that they become appreciable when the head on the weir and size of the weir decrease (see Problem 7.9, page 184).

7.8 Depth of a rectangular weir

A hydroelectric power station is located on a dammed reservoir and is largely used to generate electricity at times of peak demand. In addition to the water which passes through the turbines, a flow of water continuously bypasses the turbines and is discharged from the reservoir over a rectangular weir into the river below. When the turbines are operating at peak demand, the bypass flow is at a minimum of 34 m^{3}h^{-1} and the fall from the reservoir to the river is 5.5 m. When the power station operates at low demand, the flow of water bypassing the turbines is at a maximum of 217 m^{3}h^{-1} and the fall is 8.5 m. Determine the height of the weir crest above the surface of the river.

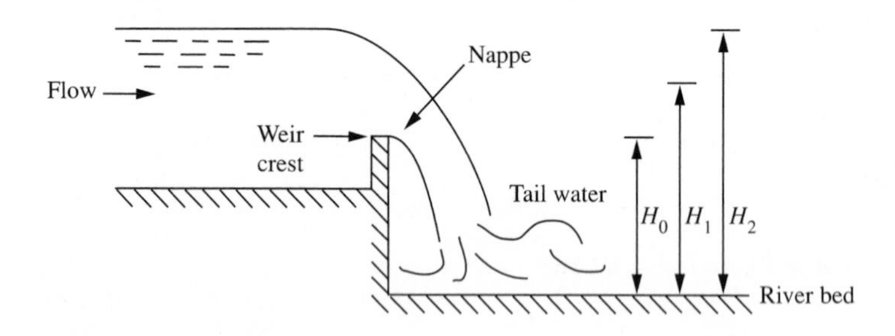

Solution

The flow over the weir of height H_1 into the river below for both cases is

$$Q_1 = \frac{2}{3} C_d B \sqrt{2g} \, (H_1 - H_0)^{\frac{3}{2}}$$

(see Problem 7.7, page 180)

and

$$Q_2 = \frac{2}{3} C_d B \sqrt{2g} \, (H_2 - H_0)^{\frac{3}{2}}$$

where B is the breadth of the weir. Assuming a constant discharge coefficient C_d then Q_1 is related to Q_2 by

$$\frac{Q_1}{Q_2} = \left(\frac{H_1 - H_0}{H_2 - H_0} \right)^{\frac{3}{2}}$$

Rearranging in terms of H_0

$$H_0 = \frac{H_2\left(\dfrac{Q_1}{Q_2}\right)^{\frac{2}{3}} - H_1}{\left(\dfrac{Q_1}{Q_2}\right)^{\frac{2}{3}} - 1}$$

$$= \frac{5.5 \times \left(\dfrac{217}{34}\right)^{\frac{2}{3}} - 8.5}{\left(\dfrac{217}{34}\right)^{\frac{2}{3}} - 1}$$

$$= 4.27 \text{ m}$$

The weir is found to be a height of 4.27 m above the river bed.

7.9 Instantaneous flow through a rectangular weir

Water used as a cooling medium in a process plant is stored in a large open tank which has a rectangular overflow weir. When the tank is overfilled the water cascades over the weir. Determine the time taken to lower the level of water to a head of 1 cm above the weir crest if it is noted that it takes 600 seconds for the level to fall from an initial head of 8 cm to 6 cm over the weir crest when the inflow to the tank has ceased.

Solution

The elemental rate of flow through the weir is

$$dQ = \frac{2}{3} C_d B \sqrt{2g} H^{\frac{3}{2}} dt$$

(see Problem 7.7, page 180)

and the discharge from the tank is

$$dQ = -AdH$$

Therefore

$$\frac{2}{3} C_d B \sqrt{2g} H^{\frac{3}{2}} dt = -AdH$$

The discharge over the weir is therefore given in general terms by

$$kH^{\frac{3}{2}} = -\frac{dH}{dt}$$

For an average head of 0.07 m then

$$k \times 0.07^{\frac{3}{2}} = -\left(\frac{-0.02}{600}\right)$$

Thus

$$k = 1.8 \times 10^{-3} \text{ m}^{-1/2}\text{s}^{-1}$$

The total time t to drain over the weir from an initial head H_1 to a final head H_2 is found by integrating with respect to H

$$\int_0^t \mathrm{d}t = -\frac{1}{k}\int_{H_1}^{H_2} H^{-\frac{3}{2}}\mathrm{d}H$$

$$= \frac{2}{k}(H_2^{-\frac{1}{2}} - H_1^{-\frac{1}{2}})$$

$$= \frac{2}{1.8\times 10^{-3}} \times (0.01^{-\frac{1}{2}} - 0.08^{-\frac{1}{2}})$$

$$= 7182 \text{ s}$$

The time is found to be approximately 2 hours. Note that it would take an infinite length of time to empty the tank, $H_2 = 0$. This equation therefore does not hold for very small values of H_2 because a layer of water adheres to the weir crest due to surface tension.

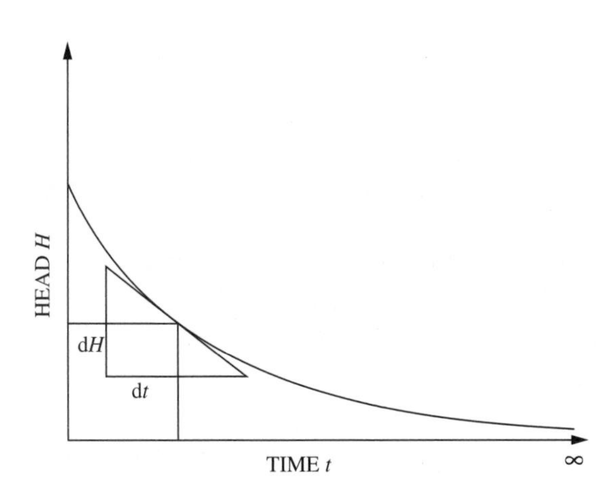

7.10 Flow through a triangular notch

A weak solution of caustic soda is fed to an absorber to remove sulphur dioxide from a process gas. The distributor to the absorber consists of channels with 90° V-notches through which the caustic soda solution discharges. Determine the rate of flow through each notch if the head above the root (bottom) is 3 cm. A discharge coefficient of 0.62 is assumed for the notches.

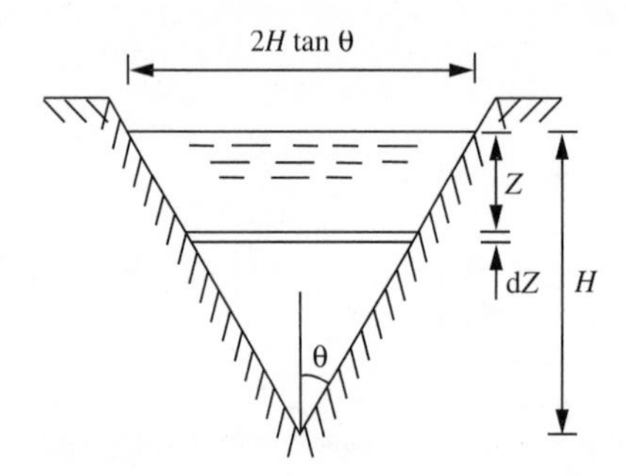

Solution

Triangular notches are generally used to measure small rates of flow in rivers and channels and can be used to channel flow for distribution purposes as illustrated in the question. For the general case for flow over a triangular notch, let the notch semi-angle be θ for which the elemental area lies at some depth between z and z+dz below the free surface. The theoretical elemental rate of flow, dQ, through the elemental area, is therefore given as

$$dQ = \sqrt{2gz}\,2(H - z)\tan\theta\, dz$$

Introducing a discharge coefficient C_d, the total actual flowrate of liquid over the notch is therefore obtained by integrating from z=0 at the free surface to z=H at the bottom (or root) of the notch. That is

$$\int_0^Q dQ = C_d 2\sqrt{2g}\tan\theta \int_0^H (Hz^{\frac{1}{2}} - z^{\frac{3}{2}})dz$$

Integrating with respect to z gives

$$Q = C_d \sqrt{2g}\, 2 \tan\theta \left[\frac{2}{3} H z^{\frac{3}{2}} - \frac{2}{5} z^{\frac{5}{2}} \right]_0^H$$

$$= \frac{8}{15} C_d \sqrt{2g} \tan\theta H^{\frac{5}{2}}$$

If the semi-angle is 45° (90° notch), the equation for liquid flowrate reduces to

$$Q = \frac{8}{15} C_d \sqrt{2g}\, H^{\frac{5}{2}}$$

$$= \frac{8}{15} \times 0.62 \times \sqrt{2g} \times 0.03^{\frac{5}{2}}$$

$$= 2.28 \times 10^{-4}\ \mathrm{m^3 s^{-1}}$$

This corresponds to a rate of flow through each notch of 0.822 m^3h^{-1}.

Note that, as with weirs and rectangular notches (see Problem 7.7, page 180) and discharging orifices (see Problem 6.2, page 145), the discharge coefficient C_d allows for energy losses and the contraction of the stream cross-section. The actual discharge through the triangular or V-notch is therefore found my multiplying the theoretical flow (or discharge) by C_d.

In this analysis the velocity of the liquid approaching the notch is assumed to be considerably less than the rate of flow through the notch. The kinetic energy of the approaching liquid is therefore neglected. That is, the velocity through the horizontal elemental area across the notch is assumed to be dependent only on the depth of flow below the free surface. If, however, the velocity of approach can not be considered to be negligible, as may be the case for a narrow channel with a single notch, the total head producing flow will be increased by the kinetic energy of the approaching liquid such that the total flow through the notch is found from

$$Q = \frac{8}{15} C_d \sqrt{2g} \tan\theta \left(\left(H + \frac{v_1^2}{2g} \right)^{\frac{5}{2}} - \left(\frac{v_1^2}{2g} \right)^{\frac{5}{2}} \right)$$

where v_1 is the average velocity of the approaching liquid in the channel.

7.11 Tank drainage through a V-notch

A sharp-edged V-notch in the side of a rectangular tank measuring 4 m long by 2 m broad gives a calibration:

$$Q = 1.5 H^{\frac{5}{2}}$$

where Q is measured in cubic metres per second and H is measured in metres. Determine the time to reduce the head in the tank from 15 cm to 5 cm if the water discharges freely through the notch and there is no inflow to the tank.

Solution

Consider the level of water to be some depth H above the bottom of the notch. A small quantity of flow dQ would reduce the depth by a level dH in time dt. Then

$$\mathrm{d}Q = 1.5 H^{\frac{5}{2}} \mathrm{d}t$$

with a change in capacity of the tank of

$$\mathrm{d}Q = -A\mathrm{d}H$$

The time to lower the level in the tank from an initial depth H_1 to a final depth H_2 is therefore found from

$$\int_0^t \mathrm{d}t = \frac{-A}{1.5} \int_{H_1}^{H_2} H^{-\frac{5}{2}} \mathrm{d}H$$

Integrating

$$t = \frac{2}{3} \frac{A}{1.5} (H_2^{-\frac{3}{2}} - H_1^{-\frac{3}{2}})$$

$$= \frac{2 \times 4 \times 2}{3 \times 1.5} \times (0.05^{-\frac{3}{2}} - 0.15^{-\frac{3}{2}})$$

$$= 257 \text{ s}$$

The time is found to be approximately 4½ minutes.

Note that the constant 1.5 is equal to

$$1.5 = \frac{8}{15} C_d \sqrt{2g}$$ (see Problem 7.10, page 186)

corresponding to a discharge coefficient C_d of 0.635.

7.12 Flow through a trapezoidal notch

Deduce an expression for the discharge through a trapezoidal notch which has a base B and a head H, and the sides of which make an angle θ to the vertical.

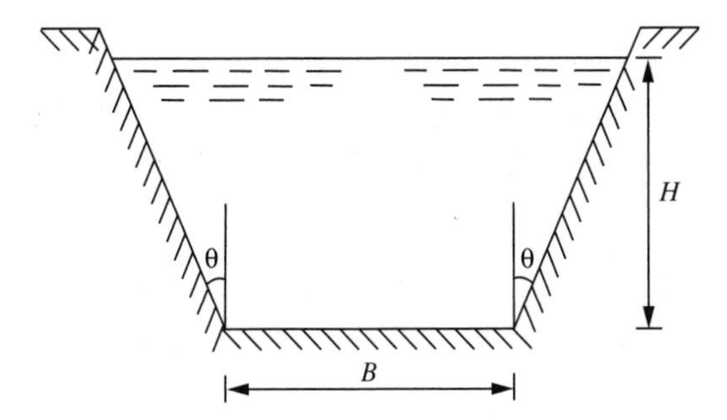

Solution

The flow can be considered as the flow through a triangular notch (see Problem 7.10, page 186) and a rectangular weir (see Problem 7.7, page 180) where the respective discharges are

$$Q_{notch} = \frac{8}{15} C_d \sqrt{2g} \tan \theta H^{\frac{5}{2}}$$

$$Q_{weir} = \frac{2}{3} C_d B \sqrt{2g}\, H^{\frac{3}{2}}$$

Combining the discharges

$$Q_{trap.notch} = \frac{8}{15} C_d \sqrt{2g} \tan \theta H^{\frac{5}{2}} + \frac{2}{3} C_d B \sqrt{2g} H^{\frac{3}{2}}$$

$$= \frac{8}{15} C_d \sqrt{2g}\, H^{\frac{3}{2}} \left(H \tan \theta + \frac{5B}{4} \right)$$

This equation may also be obtained from first principles for an elemental flow through the trapezoidal section and integrating over the total depth of flow H.

Further problems

(1) Determine the theoretical depth of liquid in a closed circular channel for maximum discharge.

Answer: 0.95 d

(2) A concrete-lined rectangular channel is to be built to transport water from a nearby river to a power station as a supply of make-up water for the cooling towers. Determine the minimum depth of the channel if the rate of flow is not expected to exceed 650 m^3min^{-1} in a channel 5 m wide with a fall of 1 in 5000. Assume the Manning formula can be used to determine the Chézy coefficient, where the roughness factor for the concrete may be taken as 0.014 $m^{-1/3}s$.

Answer: 2 m

(3) Explain why there should be a depth, less than full capacity, at which a pipe will carry water at a maximum rate of flow and suggest a way in which this can be determined analytically.

(4) Determine the angle of the walls and the relationship between breadth and depth of a channel of trapezoidal cross-section necessary to deliver maximum flow.

Answer: 30°, $B = 2/\sqrt{3H}$

(5) Water from a reservoir discharges over a rectangular weir of breadth 10 m into a sluice below. When the reservoir is at maximum capacity, the rate of discharge over the weir is 5000 m^3h^{-1} and the distance from the surface of the reservoir down to the sluice is 2 m. Determine the height of the weir crest and the discharge coefficient if the drop to the sluice is 1.9 m for a rate of discharge over the weir of 1000 m^3h^{-1}.

Answer: 1.85 m, 0.81

(6) Experimental data are recorded for the discharge through a 90° triangular notch for different heads. Plot these data on log-log paper and determine the coefficient of discharge for the notch.

Flow, m^3h^{-1}	0.30	2.99	6.95	9.70	13.02	16.94
Head, m	0.02	0.05	0.07	0.08	0.09	0.10

Answer: 0.63

(7) Derive an expression for the discharge through a weir where the area of the channel through which the water approaches the weir is such that the velocity of approach is not insignificant.

(8) Derive an expression for the rate of flow of liquid over a rectangular weir of breadth B in terms of head H.

(9) Determine the discharge coefficient for a rectangular weir placed in a channel 1 m wide of inclination 1:2500 with a Chézy coefficient of 66 $m^{1/2}s^{-1}$ if the depth in the channel is 50 cm and the height above the weir crest is 32 cm.

Answer: 0.62

(10) Determine the rate of flow through a 60° V-notch if the head of liquid above the root is 5 cm. A discharge coefficient of 0.6 may be assumed.

Answer: 1.65 m^3h^{-1}

(11) Deduce an expression for the flow of liquid through a triangular notch of angle θ.

(12) Show that the head of liquid over a weir on a sieve tray in a distillation column is given by

$$H = 0.715\left(\frac{Q}{B}\right)^{\frac{2}{3}}$$

where Q is the rate of flow (m^3s^{-1}) and B is the breadth of the weir (m).

(13) Determine the coefficient of discharge for the weir in Further Problem (12).

Answer: 0.56

(14) An aqueous solution of sodium hydroxide is fed continuously to a packed column to remove hydrogen chloride from a process gas. The liquid distributor above the packing consists of channels and the aqueous solution overflows through 60 identical 60° V-notches. Determine the head above the root of the notches in the distributor if the liquid flow to the column is 58 m^3h^{-1}. A discharge coefficient of 0.62 for the notches may be assumed.

Answer: 4 cm

Pipe friction and turbulent flow

8

Introduction

Most applications concerned with fluids in pipelines involve turbulent flow. Great efforts have been made by scientists and engineers over the years to develop empirical relationships that predict the nature of turbulent fluid flow in both rough and smooth-walled pipes. But in spite of this, no exact solutions are available which will predict precisely the nature of the turbulent flow.

French engineer Henry Philibert Gaspard Darcy in 1845 and German engineer and scientist Julius Weisbach in 1854, after much experimental work, first proposed a friction factor equation that expressed the pressure loss in a piping system in terms of velocity head. In 1911, Paul Richard Heinich Blasius, a student of Ludwig Prandtl (1875–1953), demonstrated that for smooth-walled pipes the friction factor was dependent only on the Reynolds number. Blasius produced the first plot of its kind with friction factor versus Reynolds number for the empirical relationship

$$f = 0.079\,Re^{-\frac{1}{4}}$$

valid for Reynolds numbers between 4×10^3 and 1×10^5. Three years later, British engineers Sir Thomas Ernest Stanton (a former assistant of Osborne Reynolds) and J.R. Pannell established the friction factor/Reynolds number relationship to be independent of the fluid. Over the following years several other relationships were also established of the form

$$f = a + b\,Re^{c}$$

	a	b	c	Validity
Lees (1924)	1.8×10^{-3}	0.152	−0.35	$4\times10^3<Re<4\times10^5$
Hermann (1930)	1.35×10^{-3}	0.099	−0.3	$4\times10^3<Re<2\times10^6$
Nikuradse (1932)	8×10^{-4}	0.055	−0.237	$4\times10^3<Re<3.2\times10^6$

To predict the friction of fluids in smooth pipes with turbulent flow, Prandtl developed a number of empirical models based on boundary layer theory, mixing length and wall effects. These models led to Prandtl's universal resistance equation for turbulent flow in smooth pipes given by

$$\frac{1}{\sqrt{f}} = 4\log_{10}(Re\sqrt{f}) - 0.4$$

and valid for Reynolds numbers between 5×10^3 and 3.4×10^6.

In 1858, Darcy conducted numerous detailed experiments involving pipes made from various materials. In 1932, German engineer Johannes Nikuradse analysed Darcy's data and noted that for turbulent flow in rough-walled pipes, the friction factor varied only slightly with Reynolds number, showing a decrease in friction factor with increasing Reynolds number. Above a certain limit, the friction factor was found to be independent of Reynolds number. Nikuradse also conducted his own work using artificially roughened pipes. Gluing carefully graded sand on to the inside of smooth pipes, he established a scale of relative roughness in which the diameter of sand particles in the pipe, ε, is taken as the ratio of the uncoated pipe diameter, d. Using the data of Nikuradse in 1921, the Hungarian-born aerodynamicist Theodore von Kármán (1881–1963), a co-worker of Prandtl, developed an empirical relationship for turbulent flow through rough-walled pipes in the modified form

$$\frac{1}{\sqrt{f}} = 2\log_{10}\left(\frac{d}{\varepsilon}\right) + 2.28$$

In 1939 C.F. Colebrook, working in collaboration with C.M. White, developed a mathematical function which agreed closely with values obtained for naturally rough commercial pipes. This involved using both Prandtl's smooth pipe law of friction and combining it with von Kármán's fully rough pipe law of friction into a single empirical expression in the form

$$\frac{1}{\sqrt{f}} = 2.28 - 4\log_{10}\left(\frac{\varepsilon}{d} + \frac{4.675}{Re\sqrt{f}}\right)$$

It was Louis F. Moody who, in 1944, first presented a composite plot of friction factor with Reynolds number which included the straight line relationship for laminar friction factor, smooth pipe turbulent friction factor and rough pipe turbulent friction factor, together with the concept of relative roughness. Known as the Moody plot (see page 293), it continues to be a valuable tool for evaluating friction factors for pipe flow in which it is evident that in the turbulent region, the slope of the curve relating friction factor to Reynolds number decreases with increasing Reynolds number, becoming independent of Reynolds number at high values.

Nomenclature and scientific reasoning are both major difficulties encountered in the field of pipe friction. There are several widely-accepted approaches used by industry, each having a validity which — irrespective of procedure — provides identical final answers. To the unwary, the approaches have the propensity to confuse. As a comparison, the friction factor developed in 1893 by the American engineer John Thomas Fanning (1837–1911), and known as the Fanning friction factor f, is related to other commonly-encountered forms and symbols for the friction factor where

$$f = C_f = f_F = \frac{\lambda}{4} = \frac{\phi}{4} = \frac{2\tau}{\rho v^2} = \frac{2R}{\rho v^2}$$

A common problem in calculating pressure loss through pipes in which the flow of fluid is known to be turbulent but of unknown relative roughness and/or flowrate, and thus unknown Reynolds number, is that the Fanning friction factor cannot be readily obtained. In such cases it is usual to assume either a value of 0.005 to simplify the problem, or to use this as a basis to begin an iterative procedure converging on a solution.

8.1 Economic pipe diameter

Explain what is meant by economic pipe diameter.

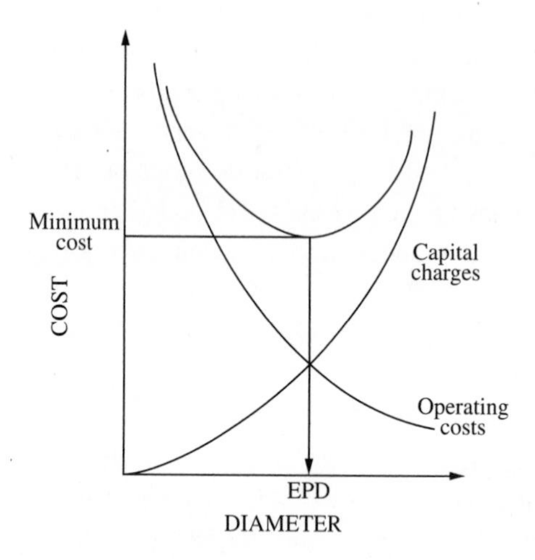

Solution

The economic pipe diameter is the diameter of pipe which gives the minimum overall cost for any specific flowrate. In general, however, pipes are determined by selecting a reasonable fluid velocity which provides a reasonable pressure drop and is virtually independent of diameter. It is possible, however, to determine the economic pipe diameter based on economic principles by which capital charges for pipes and valves are balanced against pump investment and operating costs. An economic optimum can be readily determined where the cost of pipe material appropriate to handle a particular fluid is correlated with pipe diameter. Once a calculation has been made, it is necessary to determine the Reynolds number to ensure that flow is turbulent and that other essential requirements are met, such as net positive suction head in the suction lines to centrifugal pumps.

The minimum investment is calculated for expensive or exotic pipe materials such as alloys, pipelines larger than 300 mm in diameter and carbon steel lines with a large number of valves and fittings. The pipe scheduling is selected by determining either the inner or outer diameter and the pipe wall thickness. The minimum wall thickness is a function of allowable stress of the pipe material, diameter, design pressure, and corrosion and erosion rates.

In the case of highly viscous liquids, pipelines are rarely sized on economic considerations.

8.2 Head loss due to friction

Derive an equation for the pressure and head loss due to friction for a fluid flowing through a pipe of length L and inside diameter d.

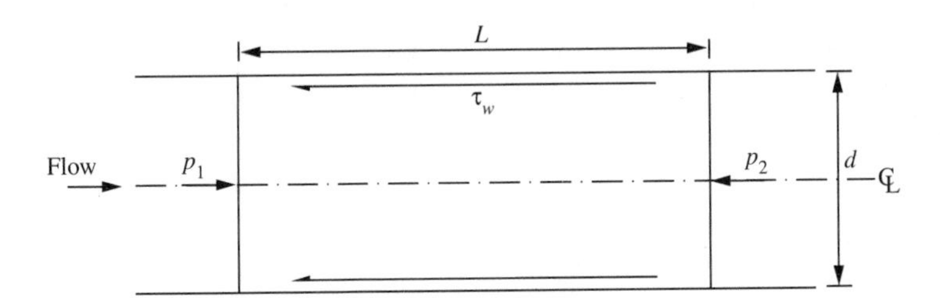

Solution

Let the pressure difference or drop be solely due to friction manifest as a wall shear stress, τ_w. For steady state conditions, a force balance on the fluid in the cross-section of the pipe is

$$p_1 \frac{\pi d^2}{4} - p_2 \frac{\pi d^2}{4} = \Delta p_f \frac{\pi d^2}{4}$$

$$= \tau_w \pi L d$$

The pressure drop due to friction is therefore

$$\Delta p_f = 4\tau_w \frac{L}{d}$$

If the wall shear stress is related to the kinetic energy per volume, then

$$\tau_w = \frac{f}{2} \rho v^2$$

where f is the Fanning friction factor. The pressure drop due to friction may therefore be expressed as

$$\Delta p_f = \frac{2 f \rho v^2 L}{d}$$

or in head form

$$H_f = \frac{4fL}{d}\frac{v^2}{2g}$$

This is known as the Fanning or Darcy equation. It was much earlier, however, that Darcy in 1845 and Weisbach in 1854 first proposed the friction factor equation after much experimental work, giving rise to the head loss due to friction in the form

$$H_f = \frac{\lambda L}{d}\frac{v^2}{2g}$$

This is known as the Darcy-Weisbach equation where the friction factor is related to the Fanning friction factor by

$$\lambda = 4f$$

8.3 General frictional pressure drop equation applied to laminar flow

Deduce a relationship between the friction factor used in the Darcy equation and Reynolds number for a fluid with fully developed laminar flow.

Solution

The average velocity of a fluid with laminar flow through the pipe is given by

$$v = \frac{1}{8\mu}\frac{\Delta p}{L}R^2$$

(see Problem 3.7, page 73)

Rearranging, the pressure drop along the pipe is therefore

$$\Delta p = \frac{8\mu L v}{R^2}$$

or in terms of pipe diameter

$$\Delta p = \frac{32\mu L v}{d^2}$$

The Darcy equation for frictional pressure drop along a pipe is

$$\Delta p_f = \frac{2f\rho v^2 L}{d}$$

Combining the two equations in terms of pressure drop

$$\frac{2f\rho v^2 L}{d} = \frac{32\mu L v}{d^2}$$

which, in terms of the Fanning friction factor, reduces to

$$f = \frac{16\mu}{\rho v d}$$

$$= \frac{16}{Re}$$

That is, for laminar flow the friction factor is inversely proportional to Reynolds number and when the log of Reynolds number is plotted against the log of the friction factor (Moody plot) gives a straight line with a gradient of -1. Note that, using the Darcy-Weisbach equation, this is equivalent to

$$\lambda = \frac{64}{Re}$$

8.4 Blasius' equation for smooth-walled pipes

Ethylbenzene, with a density of 867 kgm^{-3} and viscosity of 7.5×10^{-4} Nsm^{-2}, is to be transferred at a rate of 12 m^{3}h^{-1} through a smooth-walled pipeline 200 m long under the action of gravity. Determine the internal diameter of the pipeline if there is a fall in elevation along the pipeline of 10 m. Assume the Blasius equation can be applied to determine the friction factor.

Solution

The head loss due to friction is equal to the static head. That is

$$H = H_f$$

$$= \frac{4fL}{d}\frac{v^2}{2g}$$

where the friction factor is given by the Blasius equation

$$f = 0.079\,Re^{-\frac{1}{4}}$$

and where the Reynolds number expressed in terms of volumetric flowrate is

$$Re = \frac{4\rho Q}{\pi d\mu}$$

Therefore

$$H = \frac{4 \times 0.079\left(\dfrac{4\rho Q}{\pi d\mu}\right)^{-\frac{1}{4}} L\left(\dfrac{4Q}{\pi d^2}\right)^2}{2dg}$$

$$= \frac{4 \times 0.079 \times \left(\dfrac{4\rho Q}{\pi\mu}\right)^{-\frac{1}{4}} L\left(\dfrac{4Q}{\pi}\right)^2 d^{-\frac{19}{4}}}{2g}$$

$$10 = \frac{4 \times 0.079 \times \left(\dfrac{4 \times 867 \times \dfrac{12}{3600}}{\pi \times 7.5 \times 10^{-4}}\right)^{-\frac{1}{4}} \times 200 \times \left(\dfrac{4 \times \dfrac{12}{3600}}{\pi}\right)^2 \times d^{-\frac{19}{4}}}{2g}$$

This reduces to

$$1.44 \times 10^6 = d^{-\frac{19}{4}}$$

$$\log_e 1.44 \times 10^6 = \frac{-19}{4} \log_e d$$

$$\log_e d = -2.985$$

Solving, the diameter of the pipeline is found to be 0.0505 m (50.5 mm). A check for Reynolds number gives

$$Re = \frac{4\rho Q}{\pi d \mu}$$

$$= \frac{4 \times 867 \times \frac{12}{3600}}{\pi \times 0.0505 \times 7.5 \times 10^{-4}}$$

$$= 97{,}153$$

This value lies between 4×10^3 and 1×10^5, confirming the validity of using the Blasius equation for smooth-walled pipes.

8.5 Prandtl's universal resistance equation for smooth-walled pipes

A benzene mixture, with density of 873 kgm^{-3} and viscosity of 8.8×10^{-4} Nsm^{-2}, is fed from an open overhead tank to a process through a smooth pipe of 25 mm bore and length 18 m. If the process operates at a pressure of 55 kNm^{-2} above atmospheric, determine the minimum allowable head of the liquid surface in the tank above the process feed point if the flow is not to fall below 40 kgmin^{-1}. Allow for entrance and exit losses and apply Prandtl's universal resistance equation for turbulent flow in smooth pipes.

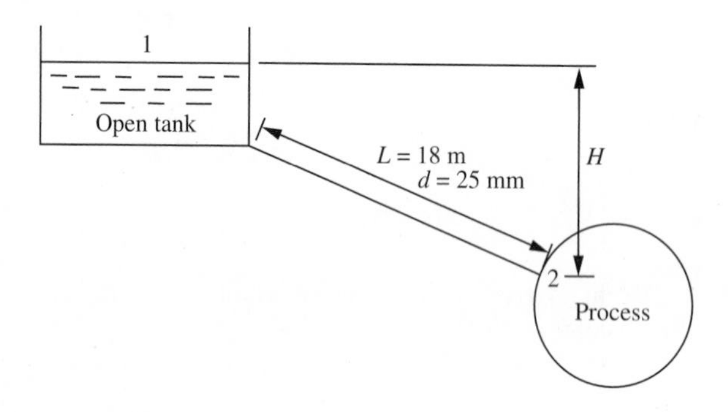

Solution

Assuming that the velocity in the overhead tank is small in comparison to the velocity in the pipe (v_1=0), then applying the Bernoulli equation between points 1 and 2

$$\frac{p_1}{\rho g} + z_1 = \frac{p_2}{\rho g} + \frac{v_2^2}{2g} + z_2 + H_L$$

The velocity v_2 can be determined from

$$v_2 = \frac{4m}{\rho \pi d^2}$$

$$= \frac{4 \times \frac{40}{60}}{873 \times \pi \times 0.025^2}$$

$$= 1.55 \text{ ms}^{-1}$$

For Reynolds number

$$Re = \frac{\rho v d}{\mu}$$

$$= \frac{873 \times 1.55 \times 0.025}{8.8 \times 10^{-4}}$$

$$= 38{,}441$$

which satisfies the criteria for Prandtl's equation ($5\times10^3<Re<3.6\times10^6$). The Prandtl equation is

$$\frac{1}{\sqrt{f}} = 4\log_{10}(Re\sqrt{f}) - 0.4$$

$$= 4\log_{10}(38{,}441\sqrt{f}) - 0.4$$

By trial and error

f	LHS	RHS
0.004	15.81	13.14
0.005	14.14	13.33
0.006	12.91	13.49
0.0055	13.48	13.42

That is, a friction factor of approximately 0.0055. The total head loss is due to pipe friction, entrance and exit losses

$$H_L = \frac{4fL}{d}\frac{v_2^2}{2g} + 0.5\frac{v_2^2}{2g} + 1.0\frac{v_2^2}{2g}$$

$$= \frac{v_2^2}{2g}\left(\frac{4fL}{d} + 1.5\right)$$

Rearranging the Bernoulli equation, the minimum allowable head is therefore

$$H = \frac{p_2 - p_1}{\rho g} + \frac{v_2^2}{2g} + \frac{v_2^2}{2g}\left(\frac{4fL}{d} + 1.5\right)$$

$$= \frac{55 \times 10^3}{873 \times g} + \frac{1.55^2}{2g} + \frac{1.55^2}{2g} \times \left(\frac{4 \times 0.0055 \times 18}{0.025} + 1.5\right)$$

$$= 8.66 \text{ m}$$

The minimum allowable height of liquid in the tank above the process feed point is 8.66 m.

8.6 Pressure drop through a rough-walled horizontal pipe

Two ethanol storage tanks are connected by 100 m of straight pipe. Both tanks are open to atmosphere and the connecting pipe has an inside diameter of 50 mm and relative roughness of 0.002. Determine the pressure drop over the length of pipe if the flowrate is 15 m^3h^{-1} and estimate the difference in level of ethanol between the two tanks. There are no extra fittings in the pipe but the entrance to the pipe and the exit from the pipe should be taken into account. The density of ethanol is 780 kgm^{-3} and its viscosity is 1.7×10^{-3} Nsm^{-2}.

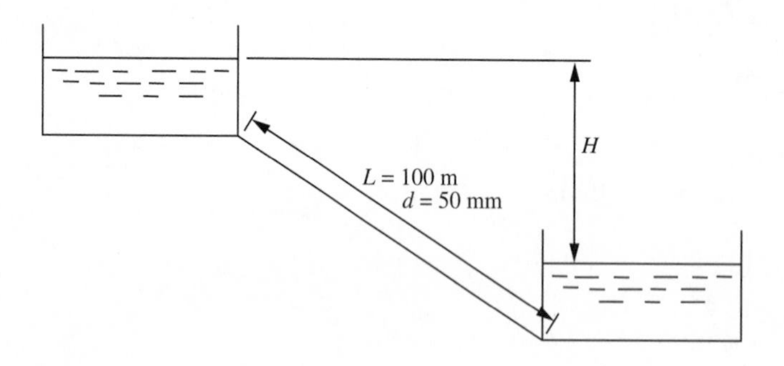

Solution

The average velocity of ethanol through the pipe is

$$v = \frac{4Q}{\pi d^2}$$

$$= \frac{4\times\frac{15}{3600}}{\pi\times 0.05^2}$$

$$= 2.12 \text{ ms}^{-1}$$

which corresponds to a Reynolds number of

$$Re = \frac{\rho v d}{\mu}$$

$$= \frac{780\times 2.12\times 0.05}{1.7\times 10^{-3}}$$

$$= 48{,}635$$

The relative roughness for the pipe, ε/d, is 0.002. From the Moody plot (see page 293) the friction factor is 0.0065. The head loss due to friction along the pipe is therefore found from

$$\Delta p_f = \frac{2f\rho v^2 L}{d}$$

$$= \frac{2 \times 0.0065 \times 780 \times 2.12^2 \times 100}{0.05}$$

$$= 91{,}146 \text{ Nm}^{-2}$$

The difference in levels between the two tanks can be determined by applying the Bernoulli equation at the free surface of both tanks from which the head loss is therefore equal to the static head. That is

$$H = H_L$$

$$= H_f + H_{exit} + H_{entrance}$$

$$= \frac{4fL}{d}\frac{v^2}{2g} + 1.0\frac{v^2}{2g} + 0.5\frac{v^2}{2g}$$

$$= \frac{v^2}{2g}\left(\frac{4fL}{d} + 1.5\right)$$

$$= \frac{2.12^2}{2g} \times \left(\frac{4 \times 0.0065 \times 100}{0.05} + 1.5\right)$$

$$= 12.25 \text{ m}$$

The pressure drop due to friction is 91.1 kNm^{-2} and the difference in levels is 12.25 m.

8.7 Discharge through a siphon

A liquid of density 1150 kgm^{-3} is siphoned from an open vat using a tube of internal diameter 25 mm and length 40 m. The tube rises vertically from its upper end a distance 10 m to the highest point. The discharge end is 2 m below the upper end and is open to atmosphere. Determine both the minimum height of liquid allowable in the vat above the open end, and the rate of flow. The minimum allowable pressure of the liquid is 18 kNm^{-2}. Assume a Fanning friction factor of 0.005.

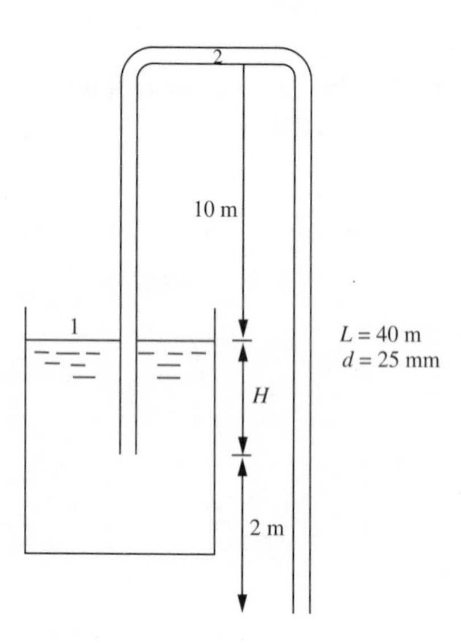

Solution

Siphoning is a useful technique for decanting a liquid from a vessel where there may be a layer of sediment which must not be disturbed. It involves the transfer of liquid to another vessel at a lower elevation by means of a pipe or tube whose highest point is above the surface of the liquid in the upper vessel. The device is started by filling the siphon tube with liquid by applying pressure on the upper surface or suction at the outlet. Liquid then continues to rise to the top point and discharge continuously under the influence of gravity to the lower vessel. A limitation, however, is that the pressure at the highest point must not fall below the vapour pressure of the liquid. Should this occur, vapour will be released from solution (boiling) and cause a break in the liquid stream. The siphoning action will also cease if the pipe entrance is no longer below the upper liquid surface. Applying the Bernoulli equation between the free surface (1) and the top point in the tube (2), then

$$\frac{p_1}{\rho g} + z_1 = \frac{p_2}{\rho g} + \frac{v_2^2}{2g} + z_2 + H_L$$

where the head loss due to friction and entrance loss is

$$H_L = \frac{4fL}{d}\frac{v^2}{2g} + 0.5\frac{v^2}{2g}$$

$$= \frac{4 \times 0.005 \times 10}{0.025} \times \frac{v^2}{2g} + 0.5 \times \frac{v^2}{2g}$$

$$= 8.5\frac{v^2}{2g}$$

The Bernoulli equation is therefore

$$\frac{101.3 \times 10^3}{1150 \times g} + H = \frac{18 \times 10^3}{1150 \times g} + \frac{v_2^2}{2g} + 10 + 8.5\frac{v^2}{2g}$$

which reduces to

$$H = 2.62 + 9.5\frac{v_2^2}{2g}$$

Applying the Bernoulli equation between the liquid surface and discharge point which are both at atmospheric pressure

$$H + 2 = \frac{v_2^2}{2g} + H_L$$

where the losses are due to tube friction, entrance and exit losses

$$H_L = \frac{4fL}{d}\frac{v^2}{2g} + 0.5\frac{v^2}{2g} + 1.0\frac{v^2}{2g}$$

$$= \frac{4 \times 0.005 \times 40}{0.025} \times \frac{v^2}{2g} + 1.5 \times \frac{v^2}{2g}$$

$$= 33.5\frac{v^2}{2g}$$

Therefore

$$2.62 + 9.5\frac{v^2}{2g} + 2 = \frac{v^2}{2g} + 33.5\frac{v^2}{2g}$$

Solving, the velocity through the tube is found to be 1.9 ms^{-1}. The corresponding flowrate is therefore

$$Q = \frac{\pi d^2}{4} v_2$$

$$= \frac{\pi \times 0.025^2}{4} \times 1.9$$

$$= 9.33 \times 10^{-4} \text{ m}^3\text{s}^{-1}$$

and the minimum depth in the vat is

$$H = 2.62 + 9.5\frac{v_2^2}{2g}$$

$$= 2.62 + 9.5 \times \frac{1.9^2}{2g}$$

$$= 4.37 \text{ m}$$

The minimum allowable height of liquid in the vat is 4.37 m. The flowrate in the siphon tube is found to be 9.33×10^{-4} m^3s^{-1}. Note that this is a problem involving variable head flow. The flowrate is therefore not constant in the tube and is at a maximum when the level in the vat is at its highest position above the open end.

8.8 Flow through parallel pipes

A liquid flows through a short pipe which branches into two parallel pipes A and B each with a length of 50 m and with inside diameters of 25 mm and 50 mm, respectively. The ends of the pipes are connected together by another short pipe. Determine the flow through each pipe if they have a drop in elevation of 3 m. Assume a constant Fanning friction factor in both pipes of 0.005.

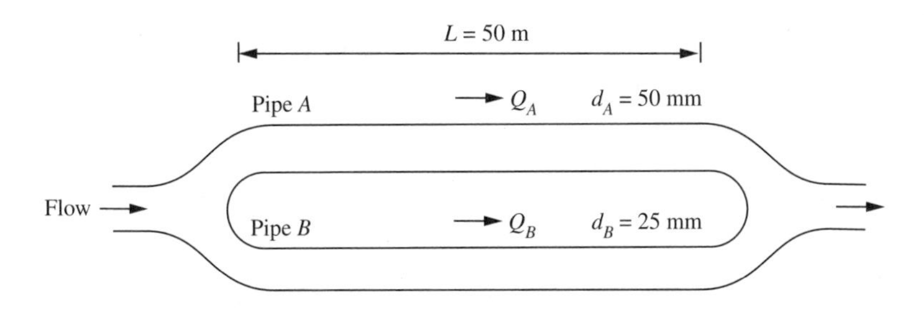

Solution

The head loss due to friction in pipe A is

$$H_f = \frac{4fL}{d}\frac{v_A^2}{2g}$$

$$= \frac{4 \times 0.005 \times 50}{0.025} \times \frac{v_A^2}{2g}$$

$$= 40\frac{v_A^2}{2g}$$

and for pipe B is

$$H_f = \frac{4fL}{d}\frac{v_B^2}{2g}$$

$$= \frac{4 \times 0.005 \times 50}{0.05} \times \frac{v_B^2}{2g}$$

$$= 20\frac{v_B^2}{2g}$$

As the pressure is the same at the end of each pipe then

$$40\frac{v_A^2}{2g} = 20\frac{v_B^2}{2g}$$

$$= H$$

$$= 3 \text{ m}$$

from which the velocities v_A and v_B are found to be 1.21 ms^{-1} and 1.71 ms^{-1} respectively. The rate of flow through the pipes is therefore

$$Q_A = \frac{\pi d_A^2}{4} v_A$$

$$= \frac{\pi \times 0.025^2}{4} \times 1.21$$

$$= 5.94 \times 10^{-4} \text{ m}^3\text{s}^{-1}$$

and

$$Q_B = \frac{\pi d_B^2}{4} v_B$$

$$= \frac{\pi \times 0.05^2}{4} \times 1.71$$

$$= 3.36 \times 10^{-3} \text{ m}^3\text{s}^{-1}$$

The rates of flow through parallel pipes A and B are therefore 2.14 m^3h^{-1} and 12.08 m^3h^{-1}, respectively, corresponding to a total flow of 14.22 m^3h^{-1}.

8.9 Pipes in series: flow by velocity head method

Two water reservoirs are connected by a straight pipe 1 km long. For the first half of its length the pipe is 12 cm in diameter after which it is suddenly reduced to 6 cm. Determine the flow through the pipe if the surface of the water in the upper reservoir is 30 m above that in the lower. Assume a friction factor of 0.005 for both pipes.

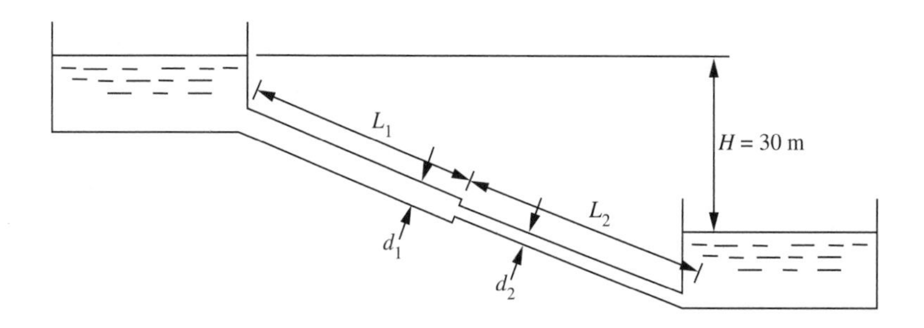

Solution

From continuity, the velocity in both sections of pipe of circular cross-section is related to flow for the incompressible fluid as

$$\frac{\pi d_1^2}{4} v_1 = \frac{\pi d_2^2}{4} v_2$$

Therefore

$$v_1 = \frac{d_2^2}{d_1^2} v_2$$

$$= \frac{0.06^2}{0.12^2} v_2$$

$$= \frac{v_2}{4}$$

Basing the calculations on the smaller diameter pipe, the head loss at the entrance to the larger pipe is

$$0.5 \frac{v_1^2}{2g} = 0.03125 \frac{v_2^2}{2g}$$ (see Problem 2.6, page 47)

The head loss in the 12 cm inside diameter pipe due to friction is given by

$$H_f = \frac{4fL_1}{d}\frac{v_1^2}{2g}$$

or in terms of the smaller 6 cm inside diameter pipe

$$= \frac{4fL_1}{16d}\frac{v_2^2}{2g}$$

$$= \frac{4 \times 0.005 \times 500}{16 \times 0.12} \times \frac{v_2^2}{2g}$$

$$= 5.2\frac{v_2^2}{2g}$$

The entrance head loss at the sudden contraction H_c between pipes is found using the chart provided in Problem 2.6 on page 47, and uses

$$H_c = k\frac{v_2^2}{2g}$$

in which k is found by interpolation to be 0.44, noting that the ratio of pipe areas a_2/a_1 is equal to 0.25. That is

$$H_c = 0.44\frac{v_2^2}{2g}$$

The head loss due to friction in the 6 cm inside diameter pipe is

$$H_f = \frac{4fL_2}{d}\frac{v_2^2}{2g}$$

$$= \frac{4 \times 0.005 \times 500}{0.06} \times \frac{v_2^2}{2g}$$

$$= 166.7\frac{v_2^2}{2g}$$

The head loss at the exit is

$$H_{exit} = 1.0\frac{v_2^2}{2g}$$ (see Problem 2.5, page 44)

The total head loss H between the two reservoirs is therefore

$$H = \frac{v_2^2}{2g} \times (0.03125 + 5.2 + 0.44 + 166.7 + 1.0)$$

$$= 173.4\frac{v_2^2}{2g}$$

This is equal to the static head between the reservoirs. That is

$$30 = 173.4\frac{v_2^2}{2g}$$

Rearranging, the velocity in the smaller pipe is therefore

$$v_2 = \sqrt{\frac{30 \times 2 \times g}{173.4}}$$

$$= 1.84 \text{ ms}^{-1}$$

The rate of flow between reservoirs is thus

$$Q = \frac{\pi d_2^2}{4} v_2$$

$$= \frac{\pi \times 0.06^2}{4} \times 1.84$$

$$= 5.2 \times 10^{-3} \text{ m}^3\text{s}^{-1}$$

The rate of flow is found to be 5.2×10^{-3} m^3s^{-1} or 1.872×10^{-2} m^3h^{-1}.

8.10 Pipes in series: pressure drop by equivalent length method

Kerosene, with density of 815 kgm^{-3} and viscosity of 7×10^{-3} Nsm^{-2}, flows through a pipe system at a rate of 12×10^3 kgh^{-1}. The pipe system consists of 50 m of 50 mm bore straight pipe with two 90° elbows, followed by a reducer to a 38 mm bore section 50 m in length with two further 90° elbows. The pipe has an absolute wall roughness of 0.02 mm. Determine the pressure loss due to friction in the system if the equivalent length of each elbow is 40 pipe diameters and the reducer is equal to 0.2 velocity heads. Assume that the von Kármán equation for rough-walled pipes applies.

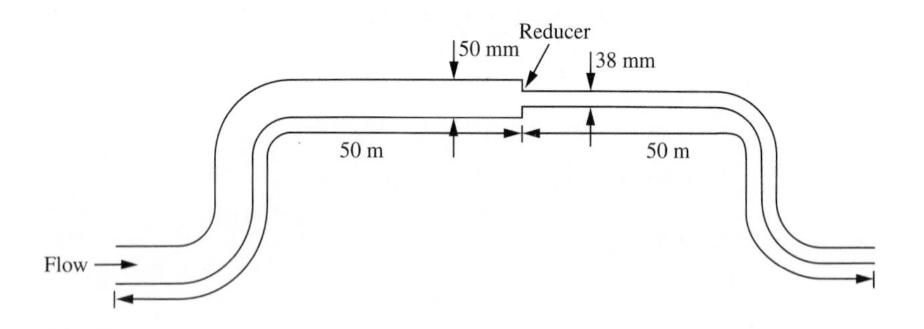

Solution

The velocity in the 50 mm bore section is

$$v = \frac{4m}{\rho\pi d}$$

$$= \frac{4\times\dfrac{12\times10^3}{3600}}{815\times\pi\times0.05^2}$$

$$= 2.08 \text{ ms}^{-1}$$

Applying the von Kármán equation

$$\frac{1}{\sqrt{f_1}} = 2\log_{10}\left(\frac{d_1}{\varepsilon}\right) + 2.28$$

$$= 2 \times \log_{10}\left(\frac{50}{0.02}\right) + 2.28$$

gives a friction factor of 0.0121. The frictional head loss through the pipe and two 90° elbows expressed in equivalent length is given by

$$H_{50\text{ mm}} = \frac{4f_1L_{eq}}{d}\frac{v^2}{2g}$$

$$= \frac{4 \times 0.0121 \times (50 + 2 \times 50 \times 0.05)}{0.05} \times \frac{2.08^2}{2g}$$

$$= 11.64 \text{ m}$$

The velocity in the 38 mm bore section is

$$v = \frac{4 \times \dfrac{12 \times 10^3}{3600}}{815 \times \pi \times 0.038^2}$$

$$= 3.61 \text{ ms}^{-1}$$

The friction factor is obtained from the von Kármán equation

$$\frac{1}{\sqrt{f_2}} = 2\log_{10}\left(\frac{d_2}{\varepsilon}\right) + 2.28$$

$$= 2 \times \log_{10}\left(\frac{38}{0.02}\right) + 2.28$$

to give a value of 0.0128. The head loss due to friction in terms of equivalent length is therefore

$$H_{38\text{ mm}} = \frac{4 \times 0.0128 \times (50 + 2 \times 50 \times 0.038)}{0.038} \times \frac{3.61^2}{2g}$$

$$= 48.15 \text{ m}$$

The head loss at the reducer, H_r, based on the smaller bore pipe is

$$H_r = 0.2\frac{v^2}{2g}$$

$$= 0.2 \times \frac{3.61^2}{2g}$$

$$= 0.13 \text{ m}$$

The total head loss is therefore

$$H_L = 11.64 + 48.15 + 0.13$$

$$= 59.92 \text{ m}$$

The pressure loss due to friction is therefore

$$\Delta p_f = \rho g H_L$$

$$= 815 \times g \times 59.92$$

$$= 479.07 \times 10^3 \text{ Nm}^{-2}$$

The total pressure drop due to friction through the pipe system is 479 kNm^{-2}.

8.11 Relationship between equivalent length and velocity head methods

Determine the equivalent friction factor relating the head loss for flow of a fluid around a 90° elbow which is expressed as both 1.2 velocity heads and 60 pipe diameters.

Solution

The head loss for the elbow expressed in terms of velocity head is

$$H_L = 1.2\frac{v^2}{2g}$$

and expressed in terms of equivalent length, the equation for frictional head loss for the elbow is

$$H_L = \frac{4fL_{eq}}{d}\frac{v^2}{2g}$$

where the equivalent length of pipe is

$$L_{eq} = 60\,d$$

Note that the equivalent length of fitting is that length of pipe which gives the same pressure drop as the fitting. Since each size of pipe (or fitting) would require a different equivalent length for a particular fitting, it is usual to express equivalent length as so many pipe diameters which is therefore independent of pipe. Therefore

$$H_L = 240f\frac{v^2}{2g}$$

Combining the two equations for head loss

$$1.2\frac{v^2}{2g} = 240f\frac{v^2}{2g}$$

Therefore

$$f = \frac{1.2}{240}$$

$$= 0.005$$

That is, the equivalent Fanning friction factor is found to be 0.005. This relationship between velocity head and equivalent length of pipe holds well for high Reynolds numbers. At low Reynolds numbers in the turbulent region the two methods deviate due to the influence of pipe surface roughness. In 1943, American hydraulics engineer Hunter Rouse gave a limiting equation that distinguished between the transition region and the fully rough regime where

$$Re = \frac{100}{\sqrt{f}\left(\frac{\varepsilon}{d}\right)}$$

Beyond the Reynolds number predicted by this equation, the friction factor becomes essentially independent of the Reynolds number. The Rouse limit line can be included on the Moody plot (see page 293).

8.12 Flow and pressure drop around a ring main

Water is supplied to a small laboratory four-side ring main ABCD at A and B, with flowrates 3×10^{-3} m^3s^{-1} and 0.5×10^{-3} m^3s^{-1}, respectively, and is drawn at rate C and D with the flowrate at D being 0.9×10^{-3} m^3s^{-1}. Determine the rate of flow from B to C and the lowest pressure drop in the main. The connecting pipes each have lengths of 10 m and internal diameters of 25 mm. The Fanning friction factor is assumed to be 0.005.

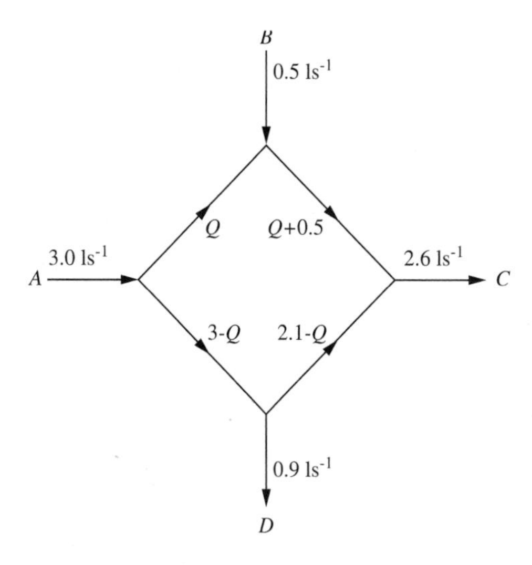

Solution

The pressure drop around the main is

$$\Delta p_{AB} + \Delta p_{BC} = \Delta p_{AD} + \Delta p_{DC}$$

where

$$\Delta p_{AD} = \frac{2f\rho v^2 L}{d}$$

$$= \frac{2f\rho L}{d}\left(\frac{4(3\times10^{-3} - Q)}{\pi d^2}\right)^2$$

$$= \frac{2\times0.005\times1000\times10\times16\times10^{-6}\times(3-Q)^2}{\pi\times0.025^5}$$

$$= 16{,}600(3-Q)^2$$

Likewise

$$\Delta p_{AB} = 16{,}600Q^2$$

$$\Delta p_{BC} = 16{,}600(Q + 0.5)^2$$

$$\Delta p_{DC} = 16{,}600(2.1 - Q)^2$$

where Q is expressed in litres per second. Therefore

$$Q^2 + (Q + 0.5)^2 = (3 - Q)^2 + (2.1 - Q)^2$$

which reduces to

$$11.2Q - 13.16 = 0$$

Solving, the flowrate Q is 1.175 litres per second. The rate of flow from B to C is therefore 1.675 litres per second. The corresponding pressure drops through the pipes are therefore

$$\Delta p_{AD} = 55.5 \text{ kNm}^{-2}$$

$$\Delta p_{AB} = 22.8 \text{ kNm}^{-2}$$

$$\Delta p_{BC} = 46.4 \text{ kNm}^{-2}$$

$$\Delta p_{DC} = 14.3 \text{ kNm}^{-2}$$

with the lowest pressure drop in the pipe connecting C and D. Note that where the friction factors, lengths and diameters are not equal, a quadratic equation in terms of flowrate arises.

8.13 Tank drainage through a pipe with turbulent flow

Water flows from an open cylindrical tank of cross-sectional area 4 m² through a 20 m length of horizontal pipe of 25 mm inside diameter. Determine the time to lower the water level in the tank from 2 m to 0.5 m above the open end of the pipe. Assume a friction factor of 0.005 and allow for entry and exit losses.

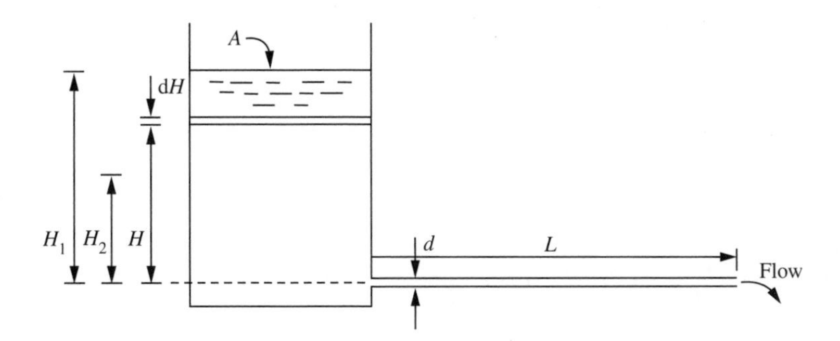

Solution

Atmospheric pressure is exerted both at the free surface in the tank and at the jet issuing from the open pipe. Applying the Bernoulli equation where it is reasonable to assume that the velocity of the water in the tank is far less than the velocity in the pipe, the static head is therefore

$$H = \frac{v^2}{2g} + H_L$$

where the head loss for pipe friction, exit and entrance loss is given by

$$H_L = \frac{4fL}{d}\frac{v^2}{2g} + 1.0\frac{v^2}{2g} + 0.5\frac{v^2}{2g}$$

$$= \frac{v^2}{2g}\left(\frac{4fL}{d} + 1.5\right)$$

Therefore

$$H = \frac{v^2}{2g} + \frac{v^2}{2g}\left(\frac{4fL}{d} + 1.5\right)$$

Rearranging

$$v = \sqrt{\frac{2gH}{\frac{4fL}{d} + 2.5}}$$

Applying an unsteady state mass balance over the tank, the change in capacity of the tank is equal to the rate of flow through the pipe. That is

$$av = -A\frac{\mathrm{d}H}{\mathrm{d}t}$$

Substituting for pipe velocity and rearranging, the total time to lower the level from H_1 to H_2 is

$$\int_0^t \mathrm{d}t = \frac{-A\sqrt{\frac{4fL}{d} + 2.5}}{a\sqrt{2g}} \int_{H_1}^{H_2} H^{-\frac{1}{2}}\mathrm{d}H$$

Integrating

$$t = \frac{2A\sqrt{\frac{4fL}{d} + 2.5}}{a\sqrt{2g}}(H_1^{\frac{1}{2}} - H_2^{\frac{1}{2}})$$

$$= \frac{2 \times 4 \times \sqrt{\frac{4 \times 0.005 \times 20}{0.025} + 2.5}}{\frac{\pi \times 0.025^2}{4} \times \sqrt{2g}} \times (2^{\frac{1}{2}} - 0.5^{\frac{1}{2}})$$

$$= 7912 \text{ s}$$

The time taken is found to be 2 hours and 12 minutes.

8.14 Turbulent flow in non-circular ducts

Hot water, at a temperature of 80°C with a corresponding density of 972 kgm^{-3} and viscosity 3.5×10^{-4} Nsm^{-2}, flows at a rate of 50 m^{3}h^{-1} through an item of process plant which consists of a horizontal annulus consisting of two concentric tubes. The outer tube has an inner diameter of 150 mm and the inner tube has an outer diameter of 100 mm. Determine the pressure loss due to friction per unit length if the surface roughness of the tubing is 0.04 mm. Use the Moody plot to obtain the friction factor.

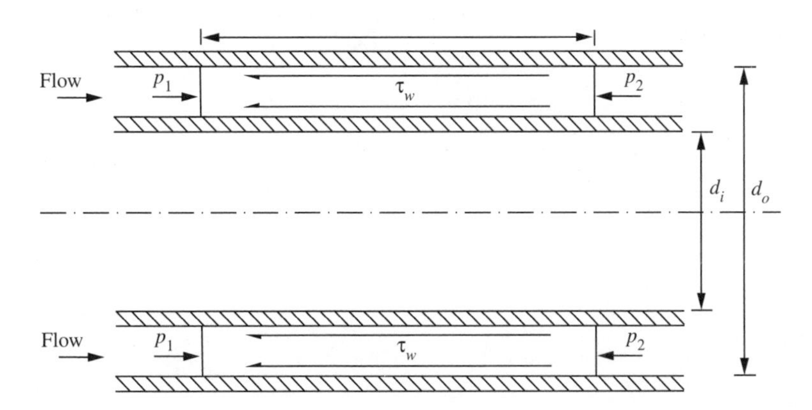

Solution

For the flow through a non-circular pipe or duct, consider a horizontal duct of any shape with uniform cross-section. A force balance on the fluid in the cross-section gives

$$(p_1 - p_2)a = \Delta p_f a$$

$$= \tau_w PL$$

where P is the wetted perimeter of the wall, L is the length of element and a is the uniform cross-sectional area. Relating the wall shear stress to friction factor in the form

$$\tau_w = \frac{f}{2}\rho v^2$$

Then

$$\Delta p_f = \frac{f\rho v^2 LP}{2a}$$

Comparing this to the equation for pressure drop due to friction in a circular pipe

$$\Delta p_f = \frac{2f\rho v^2 L}{d}$$

it is noted that $4a/P$ has replaced d. Therefore defining the term d_{eq}, known as the equivalent hydraulic diameter, as

$$d_{eq} = \frac{4a}{P}$$

the frictional pressure drop equation for non-circular pipes can be expressed in the form

$$\Delta p_f = \frac{2f\rho v^2 L}{d_{eq}}$$

The flow area for the tube annulus is

$$a = \frac{\pi}{4}(d_o^2 - d_i^2)$$

$$= \frac{\pi}{4} \times (0.15^2 - 0.1^2)$$

$$= 9.82 \times 10^{-3} \text{ m}^2$$

and the wetted perimeter is

$$P = \pi(d_o + d_i)$$

$$= \pi \times (0.15 + 0.1)$$

$$= 0.785 \text{ m}$$

The equivalent hydraulic diameter is therefore

$$d_{eq} = \frac{4 \times 9.82 \times 10^{-3}}{0.785}$$

$$= 0.05 \text{ m}$$

The relative roughness based on the equivalent hydraulic diameter is therefore

$$\frac{\varepsilon}{d} = \frac{0.04}{50}$$

$$= 8 \times 10^{-4}$$

The average velocity through the annulus is

$$v = \frac{Q}{a}$$

$$= \frac{\frac{50}{3600}}{9.82 \times 10^{-3}}$$

$$= 1.41 \text{ ms}^{-1}$$

The Reynolds number is

$$Re = \frac{\rho v d_{eq}}{\mu}$$

$$= \frac{972 \times 1.41 \times 0.05}{3.5 \times 10^{-4}}$$

$$= 195{,}788$$

From the Moody plot (page 293), the Fanning friction factor is 0.0052. The pressure loss per unit length due to friction is therefore

$$\frac{\Delta p_f}{L} = \frac{2f\rho v^2}{d_{eq}}$$

$$= \frac{2 \times 0.0052 \times 972 \times 1.41^2}{0.05}$$

$$= 402 \text{ Nm}^{-2}\text{m}^{-1}$$

That is, the pressure drop due to friction is found to be 400 Nm^{-2} per metre length of tube.

8.15 Head loss through a tapered section

Water flows through a conical section of pipe which narrows from an internal diameter of 30 cm to an internal diameter of 10 cm. The total length of the section is 3 m. Determine the frictional head loss for a flow of 0.07 m^3s^{-1} if the Fanning friction factor is 0.005. Ignore entrance and exit losses and effects due to inertia.

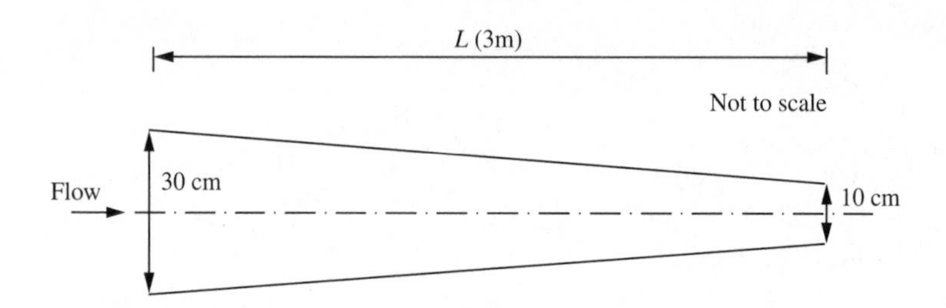

Solution

The head loss due to friction is given by

$$H_f = \frac{4fL}{d}\frac{v^2}{2g}$$

or in terms of flowrate

$$H_f = \frac{4fL}{d}\frac{\left(\frac{4Q}{\pi d^2}\right)^2}{2g}$$

$$= \frac{32fLQ^2}{\pi^2 g d^5}$$

From geometry, the diameter of the section is related to length by

$$d = 0.3 - 0.0666\,L$$

where diameter d and length L are measured in metres. The total frictional loss is therefore obtained by integration over the length of the section.

That is

$$\int_0^{H_f} \mathrm{d}H_f = \frac{32fQ^2}{\pi^2 g}\int_0^3 \frac{\mathrm{d}L}{(0.3-0.0666\,L)^5}$$

The integration is simplified using the substitution

$$u = 0.3 - 0.0666L$$

for which the differential is

$$\mathrm{d}u = -0.0666\,\mathrm{d}L$$

The integration required is therefore

$$\int_0^{H_f} \mathrm{d}H_f = \frac{-32fQ^2}{0.0666\,\pi^2 g}\int \frac{\mathrm{d}u}{u^5}$$

Thus, integrating with respect to u gives

$$H_f = \frac{-32fQ^2}{0.0666\,\pi^2 g}\frac{u^{-4}}{4}$$

That is

$$H_f = \frac{-32fQ^2}{0.0666\,\pi^2 g}\left[\frac{(0.3-0.0666\,L)^{-4}}{-4}\right]_0^3$$

$$= \frac{32\times 0.005\times 0.07^2}{0.0666\times\pi^2\times g}\times\left(\frac{(0.3-0.0666\times 3)^{-4}-0.3^{-4}}{4}\right)$$

$$= 0.3\ \mathrm{m}$$

The frictional head loss through the section is found to be 0.3 m.

8.16 Acceleration of a liquid in a pipe

Two large storage tanks containing a process liquid of density 1100 kgm^{-3} are connected by a 100 mm inside diameter pipe 50 m in length. The tanks operate with a constant difference in level of 4 m and flow is controlled by a valve located near the receiving tank. When fully open, the valve has a head loss of 13 velocity heads. If the valve is suddenly opened, determine the time taken for the process liquid to reach 99% of the final steady value. Assume a friction factor of 0.006 and neglect entrance and exit losses.

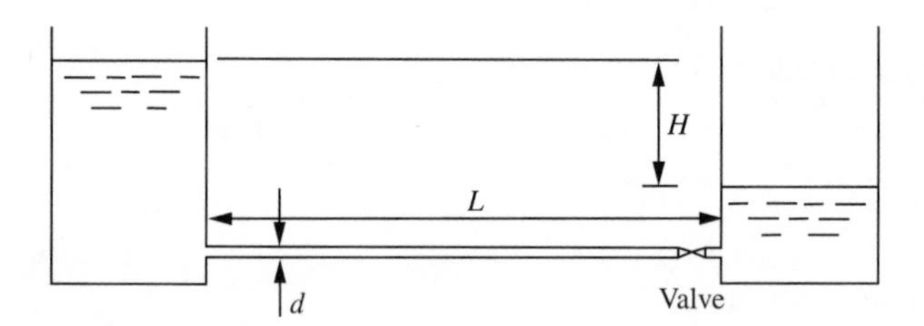

Solution

The process liquid is assumed to be inelastic so that there is no pressure wave. The average steady velocity of the liquid under full flow may be determined from the case where, for the freely discharging liquid, the static head is balanced by the frictional resistance to flow due to the pipe wall and the valve as

$$H = \frac{4fL}{d}\frac{v_1^2}{2g} + 13\frac{v_1^2}{2g}$$

Rearranging

$$v_1^2 = \frac{2gH}{\frac{4fL}{d} + 13}$$

$$= \frac{2 \times g \times 4}{\frac{4 \times 0.006 \times 50}{0.1} + 13}$$

$$= 3.14 \text{ m}^2\text{s}^{-2}$$

Therefore

$$v_1 = 1.77 \text{ ms}^{-1}$$

For the accelerating liquid, the inertia head is also included where it is assumed that the liquid is inelastic. Thus

$$H = \frac{4fL}{d}\frac{v^2}{2g} + 13\frac{v^2}{2g} + \frac{L}{g}\frac{dv}{dt}$$

Rearranging

$$\frac{2gH}{\frac{4fL}{d} + 13} = v^2 + \frac{2L}{\frac{4fL}{d} + 13}\frac{dv}{dt}$$

$$v_1^2 = v^2 + \frac{2 \times 50}{\frac{4 \times 0.006 \times 50}{0.1} + 13}\frac{dv}{dt}$$

$$= v^2 + 4\frac{dv}{dt}$$

Rearranging

$$dt = \frac{4dv}{v_1^2 - v^2}$$

$$= \frac{2}{v_1}\left(\frac{1}{v_1 + v} + \frac{1}{v_1 - v}\right)dv$$

Integrating between the limits of no velocity (stationary liquid) and 99% of the final steady value for velocity, the time taken is

$$t = \frac{2}{v_1}\left[\log_e\left(\frac{v_1 + v}{v_1 - v}\right)\right]_0^{0.99v_1}$$

$$= \frac{2}{1.77} \times \left(\log_e\left[\frac{1.77 + 0.99 \times 1.77}{1.77 - 0.99 \times 1.77}\right]\right)$$

$$= 5.98 \text{ s}$$

The time taken is found to be approximately 6 seconds for the liquid to have reached 99% of its final velocity.

Further problems

(1) Determine the pressure drop due to friction along a 1 km pipeline of inside diameter 254 mm used for drawing sea water at a rate of 500 m^3h^{-1} into a chemical plant for use as a process cooling medium. The friction factor is 0.005 and density of sea water 1021 kgm^{-3}.

Answer: 302 kNm^{-2}

(2) Determine the pressure drop due to friction along a pipe of inside diameter 100 mm, 500 m in length, carrying a process liquid of density 900 kgm^{-3} at a rate of 46.8 m^3h^{-1}. The Fanning friction factor is assumed to be 0.005.

Answer: 123.3 kNm^{-2}

(3) Using the Darcy equation, determine the pressure drop along a 10 m length of pipe of inside diameter 25.4 mm carrying a process liquid of density 1010 kgm^{-3} and viscosity 0.008 Nsm^{-2} at a rate of 0.015 m^3min^{-1}.

Answer: 1.91 kNm^{-2}

(4) Determine the pressure drop due to friction along a 1 km length of pipe of inside diameter 200 mm carrying a process liquid at a rate of 125 m^3h^{-1}. The process liquid has a density of 950 kgm^{-3} and viscosity of 9×10^{-4} Nsm^{-2} and the pipe wall has an absolute roughness of 0.04 mm.

Answer: 48.8 kNm^{-2}

(5) Determine the head loss due to friction in terms of the number of velocity heads for a flowing fluid in a pipeline of length 2.5 km and inside diameter 100 mm with a Fanning friction factor of 0.006.

Answer: $600\, v^2/2g$

(6) An experimental test rig was used to determine the parameters a and b relating the friction factor to Reynolds number in the form

$$f = a\, Re^b$$

The rig consists of 10 m of smooth-walled pipe with an inside diameter of 16 mm and uses water as the liquid medium. Results from two trials gave frictional pressure drops of 13.9 kNm^{-2} and 20.0 kNm^{-2} for flows of 0.94 m^3h^{-1} and 1.16 m^3h^{-1} , respectively. Determine the values for a and b.

Answer: $a = 0.079$, $b = -0.25$

(7) A distillation column tower operating at a gauge pressure of 700 kNm^{-2} receives a hydrocarbon feed of density 780 kgm^{-3} and viscosity 0.6×10^{-4} Nsm^{-2} at a rate of 486 m^3h^{-1} from a neighbouring column operating at a gauge pressure of 1.2 MNm^{-2}. The connecting pipework consists of 40 m of 25 cm inside diameter pipe with an absolute roughness of 0.046 mm and has a total of seven elbows and a single control valve. If the level of hydrocarbon mixture in the base of the feed column is steady at 4 m above the ground and the feed point of the receiving column is at an elevation of 15 m above the ground, determine the pressure drop across the control valve. Allow for entrance and exit losses and a loss for each bend equivalent to 0.7 velocity heads.

Answer: 387 kNm^{-2}

(8) Show that the head loss due to friction, H_f, for the flow of fluid in a circular pipe is given by

$$H_f = \frac{4fL}{d}\frac{v^2}{2g}$$

where f is the friction factor, L is the length of pipe, v is the average velocity, d is the diameter of the pipe and g the gravitational acceleration.

(9) A smooth pipe of internal diameter 150 mm is used for transporting oil of density 854 kgm^{-3} and viscosity 3.28×10^{-3} Nsm^{-2}. The pipe has a length of 1.2 km and an elevation of 15.4 m. If the absolute pressures at the lower and upper ends are 850 kNm^{-2} and 335 kNm^{-2}, respectively, determine the rate of flow assuming the Blasius equation applied for smooth pipes.

Answer: 0.0445 m^3s^{-1}

(10) Explain what is meant by the term equivalent head in the evaluation of energy losses across pipe fittings.

(11) Water at a rate of 1 m^3s^{-1} and with viscosity 1×10^{-3} Nsm^{-2} is to be transferred from an open reservoir to another one 60 m below it through a smooth pipe 2400 m long. Determine the internal diameter of the pipe using the Blasius equation for smooth pipes.

Answer: 0.48 m

(12) A pipeline, 500 m long and with an internal diameter of 150 mm, connects two large storage tanks. If the difference of liquid level in the two

tanks is 30 m, determine the free flow through the pipe. Ignore entrance and exit losses but assume a friction factor of 0.01.

Answer: 0.021 m^3s^{-1}

(13) Draw a diagram, with a brief description, to illustrate how the friction factor varies with Reynolds number.

(14) Describe the principle of the siphon and include a sketch.

(15) Explain the limiting factors of a siphon used to transfer a liquid.

(16) A four-side ring main *ABDC* is supplied with water at *A* for distribution in a process plant at points *B*, *C* and *D* at rates of 0.25 m^3s^{-1}, 0.1 m^3s^{-1} and 0.05 m^3s^{-1}. Determine the flowrate in pipe *AB* if the pipe details are:

Pipe	Length, m	Diameter, m
AB	1000	0.5
BC	1500	0.3
CD	500	0.3
DA	1000	0.4

The Fanning friction factor is assumed to be 0.005 in each of the pipes.

Answer: 0.274 m^3s^{-1}

(17) Determine the pressure drop along a 40 m length of pipe in terms of velocity head and equivalent length approaches for the flow of water at a rate of 20 m^3h^{-1}, if the pipe contains a gate valve (open) and three 90° elbows. The internal diameter of the pipe is 100 mm, the velocity head values for the valve and elbows are 0.15 and 0.7 velocity heads and the equivalent length of pipe are 7 and 35 pipe diameters, respectively. The Fanning friction factor is 0.006.

Answer: 2.96 kNm^{-2} and 3.07 kNm^{-2}

(18) Explain why calculations of the pressure drop due to friction for flow of a liquid through a pipe using the velocity head and equivalent length of pipe approaches do not necessarily provide the same result.

(19) Water is supplied to a three-side ring main *ABC* with a flow of 0.1 m^3s^{-1} at *A* and is drawn from *B* and *C* at rates of 0.06 m^3s^{-1} and 0.04 m^3s^{-1}, respectively. The internal diameters of the pipes are: *AB* 50 mm, *AC* 38 mm and *BC* 38

mm, and each has a length of 50 m. Determine the rate of flow in pipe *BC* if the friction factor is the same and constant for each pipe.

Answer: 0.007 m^3s^{-1}

(20) If, in the previous problem, the friction factor is given by the Blasius formula for smooth-walled pipes ($f = 0.079Re^{-1/4}$), comment on the effect this would have and determine the flow in pipe *BC*.

Answer: 0.0061 m^3s^{-1}

(21) A fermented broth of density 990 kgm^{-3} is decanted from a fermentation vessel to a receiving vessel by siphon. The siphon tube is 10 m long with a bore of 25.4 mm. Determine the rate of transfer if the difference in head between the two vessels is 2 m. Determine the pressure at the highest point of the siphon tube which is 1.5 m above the broth level and the tube length to that point is 4 m. Assume a friction factor of 0.004 and standard atmospheric pressure.

Answer: 1.136×10^{-3} m^3s^{-1}, 74.3 kNm^{-2}

(22) A 0.155 m internal diameter pipeline is fed with water from a constant head tank system. The water is discharged from the pipeline 12.5 m below the level of the water in the head tank. The pipeline contains four 90° elbows, a fully open globe valve, a half open gate valve and a three quarters open gate valve. The respective equivalent length for elbow, open, half open and three quarters open valves are 40, 300, 200 and 40 pipe diameters, respectively. Determine the rate of flow if the Fanning friction factor is 0.025.

Answer: 3.018×10^{-2} m^3s^{-1}

(23) A cylindrical tank with a diameter of 1.5 m mounted on its axis contains a liquid at a depth of 2 m. The liquid is discharged from the bottom of the tank through a 10 m length of pipe of internal diameter 50 mm at an elevation 1 m below the bottom of the tank. Flow is regulated by a valve. Determine the time to discharge half the liquid if the valve is open such that the energy loss through the valve is equivalent to 40 pipe diameters. The Fanning friction factor is assumed to be 0.008.

Answer: 412 s

Pumps

9

Introduction

Pumps are machines for transporting fluids from one place to another, usually along pipelines. They are classified as centrifugal, axial, reciprocating and rotary and may be further grouped as either dynamic or positive displacement pumps. Dynamic pumps, which include centrifugal and axial pumps, operate by developing a high liquid velocity (kinetic energy) and converting it into pressure. To produce high rates of discharge, dynamic pumps operate at high speeds, although their optimal efficiency tends to be limited to a narrow range of flows. Positive displacement pumps operate by drawing liquid into a chamber or cylinder by the action of a piston; the liquid is then discharged in the required direction by the use of check valves. This results in a pulsed flow. Positive displacement pumps, however, are capable of delivering significantly higher heads than dynamic pumps. Rotary pumps are another form of positive displacement pump capable of delivering high heads; in this case fluids are transported between the teeth of rotating and closely meshing gears or rotors and the pump casing or stator. Unlike reciprocating pumps the flow is continuous, although rotary pumps tend to operate at lower speeds than dynamic pumps and their physical size is likely to be larger.

In the decision-making process for the specification and selection of a pump to provide safe, reliable and efficient operation for a particular application, the general procedure follows many well-defined steps. Flow regulation and head requirements are the two most obvious fundamental factors that must be considered. Equally, however, the physical characteristics of the fluids — viscosity and lubrication properties, solids content and abrasiveness, as well as corrosion and erosion properties — significantly influence the choice of pump type. A complete analysis of the system to discover the requirement for the number of pumps and their individual flow is also necessary. This analysis must include basic calculations for differential static head to provide an estimate of the necessary pump head.

Once an appropriate type of pump has been identified for a particular fluid, it is then necessary to establish the relationship between head and flowrate that can be delivered by the pump to match the system requirements in terms of static head and frictional losses. There are many additional calculations and checks that must be made before the analysis is complete. It is essential, for example, to ensure that pumps are correctly located to ensure avoidance of undesirable effects such as cavitation in centrifugal pumps and separation in reciprocating pumps. Clearly, not all pumping installations demand the same level of attention. Pumps required to deliver clean, cold water with modest delivery heads, for example, are less complex than those whose duties involve high temperature and pressure, high viscosity and abrasive liquids.

In practice, there are numerous additional and important considerations that must be taken into account to ensure the safe, reliable and efficient installation and operation of a pump. It is important to consider maintenance requirements, implications and likelihood of leaks, availability of spares, economic factors such as capital and operating costs, safety for personnel and environmental considerations including noise. For the engineer new to pump specification and selection, the decision-making process may appear complicated, confusing and even, at times, conflicting. Experience and familiarity with pump selection eventually results in confidence, thereby reducing the effort required in the successful procurement, installation and operation of a pump.

9.1 Centrifugal pumps

Describe the essential features and merits of centrifugal pumps and possible reasons for vibration during operation.

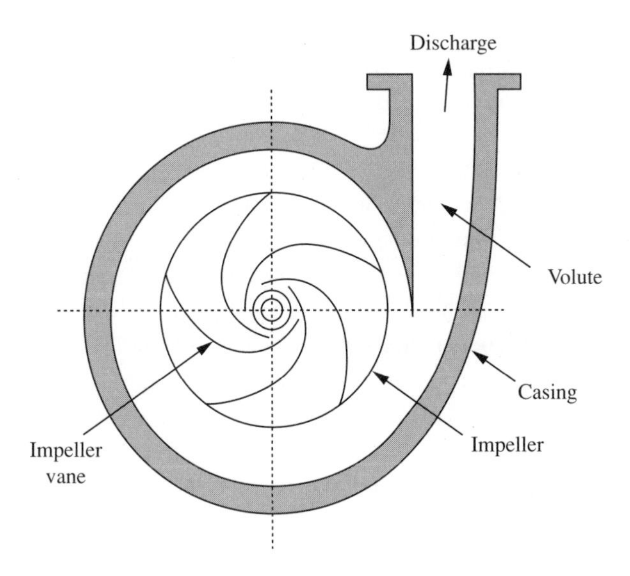

Solution

Centrifugal pumps are by far the most commonly used type of pump in the process industries, because they are versatile and relatively inexpensive. They are very suitable for handling suspended solids, able to continue operating when the delivery line is blocked, have low maintenance costs and are easily fabricated in a wide range of corrosion-resistant materials. However, they are unable to develop high heads unless multiple stages are used and they also require priming by ancillary equipment. They offer reasonable efficiency over only a limited range of conditions and are not particularly suitable for very viscous fluids.

The design of a centrifugal pump consists of a series of blades attached at the centre of a shaft known as an impeller and rotated at high speed inside a casing. Fluid is fed in axially at the centre (eye) of the impeller and is thrown out in a roughly radial direction by the centrifugal action. The large increase in kinetic energy which results is converted into pressure energy at the pump outlet either by using a volute chamber or a diffuser; the latter is more efficient but more expensive. There are considerable variations in impeller design, but almost all have blades which are curved, usually backward to the direction of rotation.

This arrangement gives the most stable flow characteristic. Centrifugal pumps do not operate by positive displacement and it is important to note that the head they develop depends not only on the size and rotational speed of the impeller but also on the volumetric flowrate.

There are, in fact, many different types of centrifugal pump. Each is intended to perform a specific duty, such as handling various types of liquids including slurries, operating at high temperature or delivering high heads or flow rates. A variation of the single impeller pump shown is the two-stage centrifugal pump. This pump has two impellers mounted on the shaft such that the output from the first impeller is fed into the second impeller. By operating impellers in series, higher heads can be developed. Multi-stage centrifugal pumps have three or more impellers mounted in series on the same shaft and these pumps are therefore capable of producing considerably higher heads.

In general, centrifugal pumps can be classified according to their construction and layout in terms of impeller suction, type of volute, nozzle location, shaft orientation, bearing support and coupling to the driver. Single suction pumps, for example, have a suction cavity on one side of the impeller only, whereas double suction pumps have suction cavities at either side of the impeller. Single suction pumps, however, are subject to higher axial thrust imbalance due to flow coming in on one side of the impeller only. Pumps with a horizontal shaft are popular due to ease of servicing and maintenance, although pumps with a shaft in the vertical plane tend to be used where space is perhaps limited. Shafts that are unsupported by a bearing are known as *overhung* and can be fed exactly at the eye of the impeller. Shafts with bearing support on both ends such that the impeller is located in between the bearings, however, provide less shaft deflection although the shaft actually blocks the impeller eye.

It is not uncommon for centrifugal pumps to be troubled by vibration during operation. This may be due to the destructive phenomenon of cavitation in which vapour is released from solution leading to a loss of discharge and delivered head. The source of the vibration may, however, be due to operating the pump dry, against a blocked or closed delivery line. It may also be due to the shaft being out of alignment, an unbalanced impeller, bearing or seal wear, or the pump being incorrectly wired. Vibration may also be due to poor pipe support, the effects of solids, including foreign bodies, in the fluid, and air locks.

9.2 Centrifugal pump matching

A centrifugal pump is used to transfer a liquid between two open storage tanks. A recycle loop mixes the contents of the feed tank and a restriction orifice in the recycle line is used to limit the liquid velocity to 2 ms^{-1}. Determine the number of velocity heads across the restriction orifice when a valve in the transfer line to the receiving tank is closed. If the valve is fully open, show that the flowrate is approximately 1.1×10^{-2} m^{3}s^{-1} when the level of the liquid in the receiving tank is 12 m above the centre line of the pump and the level in the feed tank is 5 m above the centre line. The pump characteristic is

$$H = 12 - 70Q - 4300Q^2$$

All pipelines have an inside diameter of 100 mm and the suction, recycle and transfer lines have equivalent lengths of 2 m, 10 m and 20 m, respectively. A Fanning friction factor of 0.005 may be assumed.

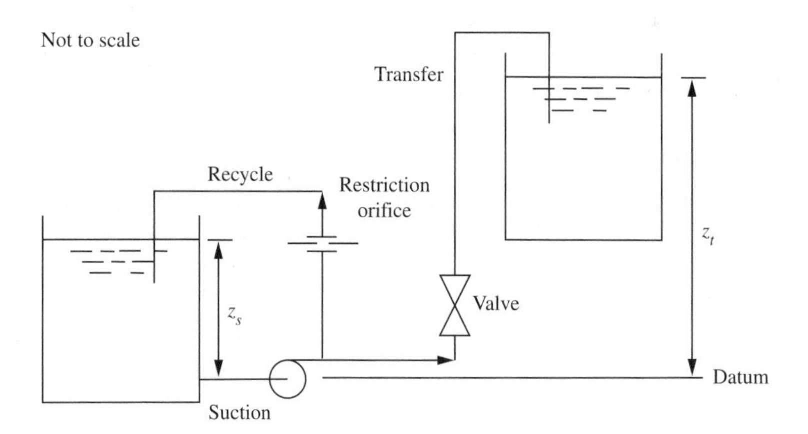

Solution

To determine the head loss across the restriction orifice consider the case in which the valve is closed. The rate of flow through the recycle line is therefore

$$Q = \frac{\pi d^2}{4} v_r$$

$$= \frac{\pi \times 0.1^2}{4} \times 2$$

$$= 0.0157 \text{ m}^3\text{s}^{-1}$$

where v_r is the average velocity in the recycle line limited to 2 ms^{-1} by the restriction orifice.

From the pump characteristic, the delivered head by the centrifugal pump is thus

$$H = 12 - 70Q - 4300Q^2$$
$$= 12 - 70 \times 0.0157 - 4300 \times 0.0157^2$$
$$= 9.84 \text{ m}$$

At the duty point this is equal to the system characteristic. Applying the Bernoulli equation to the suction and recycle lines, the head loss is due to friction, loss through the restriction orifice and exit loss from the pipe. That is

$$H = \frac{4f(L_s + L_r)}{d}\frac{v_r^2}{2g} + k\frac{v_r^2}{2g} + 1.0\frac{v_r^2}{2g}$$

where L_s and L_r are the suction and recycle equivalent lengths of pipes, respectively. Rearranging

$$k = \frac{2gH}{v_r^2} - \frac{4f(L_s + L_r)}{d} - 1$$
$$= \frac{2 \times g \times 9.84}{2^2} - \frac{4 \times 0.005 \times (2 + 10)}{0.1} - 1$$
$$= 44.9$$

That is, the head loss across the restriction orifice is equal to 44.9 velocity heads.

For the case of the valve, being fully open allowing transfer of liquid to the receiving tank, the velocity in the transfer line, v_t, is

$$v_t = \frac{4Q_t}{\pi d^2}$$
$$= \frac{4 \times 1.1 \times 10^{-2}}{\pi \times 0.1^2}$$
$$= 1.4 \text{ ms}^{-1}$$

where Q_t is the rate of transfer. Applying the Bernoulli equation to both suction and recycle lines, the system head is equal to the frictional head loss in the recycle line, suction line, across the restriction orifice and pipe exit. That is

$$H = \frac{4fL_r}{d}\frac{v_r^2}{2g} + \frac{4fL_s}{d}\frac{(v_r^2 + v_t^2)}{2g} + k\frac{v_r^2}{2g} + 1.0\frac{v_r^2}{2g}$$

Likewise, applying the Bernoulli equation to both the suction and transfer line, the system head is equal to the frictional head loss in the suction line, transfer line and pipe exit loss as well as the difference in static head between the liquid levels in the two tanks. That is

$$H = z_t - z_s + \frac{4fL_s}{d}\frac{(v_r^2 + v_t^2)}{2g} + \frac{4fL_t}{d}\frac{v_t^2}{2g} + 1.0\frac{v_t^2}{2g}$$

Equating and rearranging, the velocity in the recycle line is therefore

$$v_r = \left(\frac{2g(z_t - z_s) + \left(\frac{4fL_t}{d} + 1 \right) v_t^2}{\frac{4fL_r}{d} + k + 1} \right)^{\frac{1}{2}}$$

$$= \left(\frac{2 \times g \times (12 - 5) + \left(\frac{4 \times 0.005 \times 20}{0.1} + 1 \right) \times 1.4^2}{\frac{4 \times 0.005 \times 10}{0.1} + 44.9 + 1} \right)^{\frac{1}{2}}$$

$$= 1.75 \text{ ms}^{-1}$$

This therefore corresponds to a head of

$$H = \frac{4fL_r}{d}\frac{v_r^2}{2g} + \frac{4fL_s}{d}\frac{(v_r^2 + v_t^2)}{2g} + k\frac{v_r^2}{2g} + 1.0\frac{v_r^2}{2g}$$

$$= \frac{4fL_s}{d}\frac{v_t^2}{2g} + \left(\frac{4fL_r + L_s}{d} + k + 1 \right)\frac{v_r^2}{2g}$$

$$= \frac{4 \times 0.005 \times 2}{0.1} \times \frac{1.4^2}{2g} + \left(\frac{4 \times 0.005 \times (10 + 2)}{0.1} + 44.9 + 1 \right) \times \frac{1.75^2}{2g}$$

$$= 7.49 \text{ m}$$

The total flowrate delivered by the pump (through the transfer and recycle lines) is therefore

$$Q = Q_t + Q_r$$

$$= \frac{\pi d^2}{4}(v_t + v_r)$$

$$= \frac{\pi \times 0.1^2}{4} \times (1.75 + 1.4)$$

$$= 0.0247 \text{ m}^3\text{s}^{-1}$$

which, from the pump characteristic equation, corresponds to a delivered pump head of

$$H = 12 - 70Q - 4300Q^2$$

$$= 12 - 70 \times 0.0247 - 4300 \times 0.0247^2$$

$$= 7.65 \text{ m}$$

The delivered head therefore closely matches the system head at the flowrate of $1.1{\times}10^{-2}$ m^2s^{-1}, corresponding to the duty point.

It should be noted that while the duty point corresponds to the maximum flow of liquid delivered by the pump for the system illustrated, it may not correspond to the most efficient operation of the pump (best efficiency point, or 'bep'). Also, to ensure that cavitation is avoided, details of the pump's required net positive suction head (NPSHR) are also required (see Problem 9.5, page 245).

9.3 Centrifugal pumps in series and parallel

Highlight the benefits of using centrifugal pumps in series and parallel.

Solution

There are occasions when it is desirable to use centrifugal pumps to deliver a higher head than is normally possible for a fixed flowrate. This can be achieved by using two pumps in series. The total pump characteristic is the sum of the individual characteristics. Pumps in series are often used for high pressure applications such as boiler feed pumps. Special designs are often used — for example, running the impellers on a common shaft with the casing laid out so that the volute of one pump leads into the eye of the next.

It is possible to deliver high flowrates for a given head by using similar pumps in a parallel arrangement where the pump characteristics are added together.

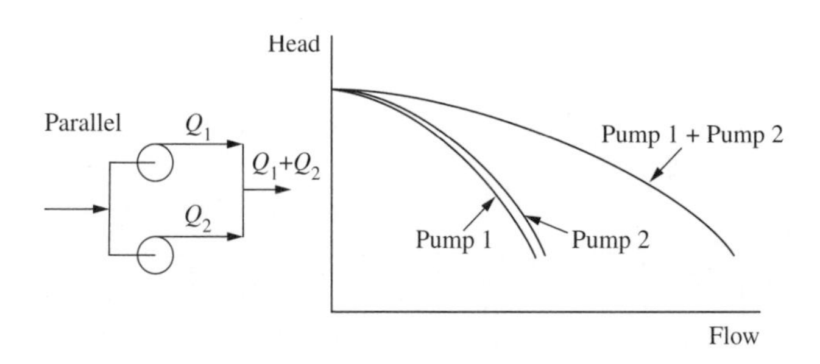

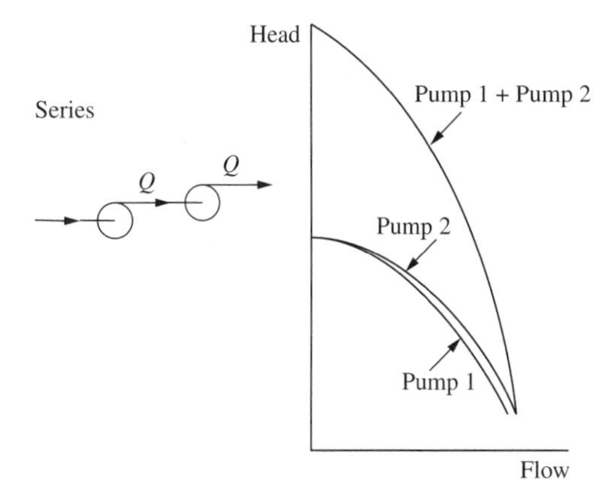

9.4 Cavitation in centrifugal pumps

Explain what is meant by cavitation in centrifugal pumps.

Solution

Vapour forms in any liquid when the pressure in the liquid is less than the vapour pressure at the liquid temperature. The possibility of this happening is much greater when the liquid is in motion, particularly on the suction side of centrifugal pumps where velocities may be high and the pressure correspondingly reduced. Having been formed, the vapour bubbles travel with the liquid and eventually collapse with explosive force giving pressure waves of high intensity. This collapse, or cavitation, occurs on the impeller blades causing noise as well as vibration and erosion of the blades, which may eventually result in a typically pitted appearance similar to that of corrosion. Apart from the audible and destructive effect of cavitation, another sign is a rapid decrease in delivered head and pump efficiency. As a remedy, a throttle valve should be placed in the delivery line and when any symptoms of cavitation are observed the valve can be partially closed, thereby restricting the throughput and reducing the kinetic energy.

Cavitation is more likely to occur with high-speed pumps, hot liquids and liquids with a high volatility (high vapour pressure). Problems can also be encountered with slurries and liquids with dissolved gas, and in particular in gas scrubbers which usually operate with liquids saturated with gas. Note that the related phenomenon of separation can occur in reciprocating pumps due to the acceleration of the liquid. This is more likely when the delivery lines are long and the corresponding kinetic energy is large. Apart from providing sufficient net positive suction head (see Problem 9.5, page 245), separation can be avoided by attaching air vessels to both the delivery and suction lines.

9.5 Net positive suction head: definition

Explain what is meant by available and required net positive suction head (NPSH).

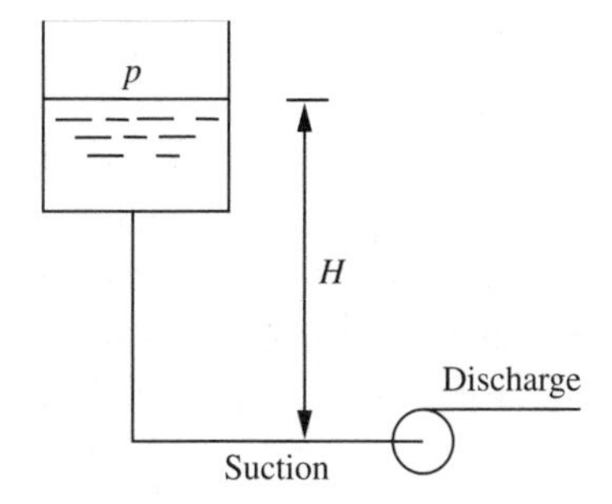

Solution

To avoid problems of cavitation in centrifugal pumps it is necessary for the lowest pressure of the liquid in the pump, which is usually at the eye of the impeller, to exceed the vapour pressure of the liquid being delivered. Should the pressure fall below the vapour pressure of the liquid, local vaporization (boiling) is likely to occur although the precise mechanism of cavitation inception is not fully understood. Since the vapour pressure of liquids is a function of temperature, it may be the case that cavitation is more likely to occur during the hot months of summer than in cold winters. Atmospheric conditions may also affect the likelihood of cavitation occurring at low barometric pressures. The required net positive suction head (NPSH) or NPSHR is therefore the positive head necessary to overcome the pressure drop in the pump and maintain the liquid above its vapour pressure. It is a function of pump design, impeller size and speed, and the rate of delivery. To prevent cavitation from occurring this must be exceeded by the available NPSH or NPSHA and is a function of the system in which the pump operates. This includes the minimum working gauge pressure of the vapour on the surface of the liquids p, the minimum atmospheric pressure taken as 0.94 of standard atmospheric pressure p_a, the vapour pressure of the liquid at the maximum operating temperature p_v, the static head above the pump centre H, and the head loss due to friction and fittings H_L. The available NPSH or NPSHA at the pump suction is then

$$\text{NPSHA} = \frac{p + p_a - p_v}{\rho g} + H - H_L$$

and is effectively a measure of the total head available at the pump suction above the vapour pressure. To ensure that vapour bubble formation at the eye of the impeller — and thus problems of cavitation — are avoided, it is essential that the NPSHA is greater than the NPSHR. That is

$$\text{NPSHA} > \text{NPSHR}$$

In the case of reciprocating type positive displacement pumps, the same approach to calculating the available NPSH is used. This allows for the static liquid head at the pump inlet, minimum barometric pressure, minimum working pressure and vapour pressure of the liquid at the maximum operating temperature. It should be noted, however, that the maximum frictional head in the pump suction line occurs at the point of peak instantaneous flow and not the average flow. This is dependent on the configuration of the pump and has greatest effect for single cylinder (or simplex) pumps (see Problem 9.15, page 265). Additionally, a head correction is required to allow for the acceleration of the fluid in the suction line during each pulsation cycle.

As with reciprocating pumps, the available NPSH for rotary pumps which operate with pulsed flow requires correction for both frictional head loss in the suction line at peak instantaneous flow and head required to accelerate the fluid in the suction line during each pulsation cycle.

Note that while the available NPSH is a function of the system in which the pump operates, the vapour pressure of the liquid being pumped is not always specifically included in the NSPHA calculation. In this case the vapour pressure head is included such that

$$\text{NPSHA} > \text{NPSHR} + \frac{p_v}{\rho g}$$

Details of the required NPSH are usually supplied by the pump manufacturer.

9.6 Net positive suction head: calculation 1

A centrifugal pump is used to deliver a liquid of density 970 kgm^{-3} from an open storage vessel at a rate of 5 m^{3}h^{-1}. The storage vessel has a diameter of 3 m and is initially at a depth of 2.5 m. The pump is located at an elevation of 3 m above the bottom of the vessel and the frictional head loss in the suction pipe is 0.5 m. The vapour pressure of the liquid at the temperature of operation is 18 kNm^{-2} and the NPSH is 5 m. Determine the quantity of liquid delivered and the time taken before cavitation occurs. Allow for worst case meteorological conditions.

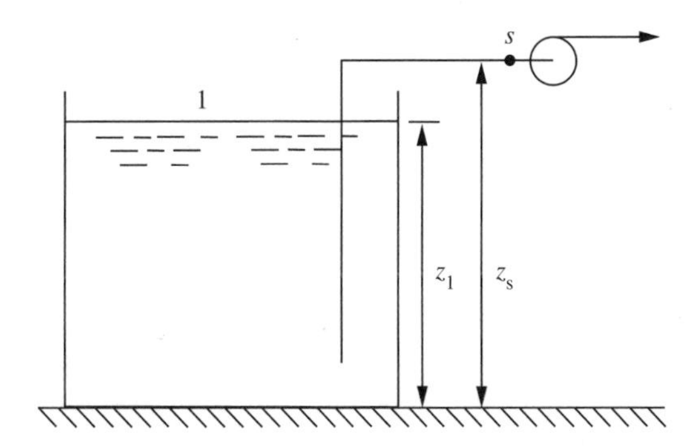

Solution

Applying the Bernoulli equation between the free surface of the liquid (1) and the suction point (subscript *s*) in the pump

$$\frac{p_1}{\rho g} + z_1 = \frac{p_s}{\rho g} + \frac{v_s^2}{2g} + z_s + H_L$$

Rearranging, the suction pressure head is therefore

$$\frac{p_s}{\rho g} = \frac{p_1}{\rho g} + z_1 - z_s - \frac{v_s^2}{2g} - H_L$$

The minimum suction head before cavitation occurs is

$$\frac{p_s}{\rho g} = \text{NPSH} - \frac{v_s^2}{2g} + \frac{p_v}{\rho g}$$

Thus, for cavitation not to occur

$$\text{NPSH} - \frac{v_s^2}{2g} + \frac{p_v}{\rho g} = \frac{p_1}{\rho g} + z_1 - z_s - \frac{v_s^2}{2g} - H_L$$

Rearranging, and noting that the pressure on the free surface is taken as 0.94 of the standard atmospheric pressure, the depth that can be pumped before cavitation occurs is

$$z_1 = \text{NPSH} + \frac{p_v - p_1}{\rho g} + z_s + H_L$$

$$= 5 + \frac{(18 - 0.94 \times 101.3) \times 10^3}{970 \times g} + 3 + 0.5$$

$$= 0.375 \text{ m}$$

corresponding to a volume of

$$V = \frac{\pi}{4} \times 3^2 \times (2.5 - 0.375)$$

$$= 15.02 \text{ m}^2$$

The time taken to reach this level is therefore

$$t = \frac{V}{Q}$$

$$= \frac{15.02}{5}$$

$$= 3 \text{ h}$$

The quantity of liquid delivered is therefore 15 m^3, taking 3 hours before cavitation occurs.

9.7 Specific speed

Explain what is meant by specific speed of a centrifugal pump.

Solution

The specific speed is useful for the selection and scale-up of centrifugal pumps and is a measure of pump performance in terms of pump discharge Q, head H and impeller speed N. It is always evaluated at the point of maximum efficiency, otherwise known as the best efficiency point. The general procedure is that once the head and the flow are established for a particular duty, the pump's specific speed can be determined to ensure the selection of the appropriate pump in terms of impeller type and characteristics with optimal operating efficiency. This is determined for a head coefficient C_H and capacity coefficient C_Q, which, for two geometrically similar pumps, are

$$C_H = \frac{gH_1}{N_1^2 D_1^2} = \frac{gH_2}{N_2^2 D_2^2}$$

(see Problem 4.3, page 104)

and

$$C_Q = \frac{Q_1}{N_1 D_1^3} = \frac{Q_2}{N_2 D_2^3}$$

From the head coefficients, the ratio of pump impellers is

$$\frac{D_1}{D_2} = \frac{N_2}{N_1}\left(\frac{H_1}{H_2}\right)^{\frac{1}{2}}$$

and from the capacity coefficients, the ratio of discharges is

$$\frac{Q_1}{Q_2} = \frac{N_1}{N_2}\left(\frac{D_1}{D_2}\right)^3$$

The ratio of discharges is therefore

$$\frac{Q_1}{Q_2} = \frac{N_1}{N_2}\left(\frac{N_2}{N_1}\left(\frac{H_1}{H_2}\right)^{\frac{1}{2}}\right)^3$$

$$= \left(\frac{N_2}{N_1}\right)^2\left(\frac{H_1}{H_2}\right)^{\frac{3}{2}}$$

or rearranging

$$\frac{N_1 Q_1^{\frac{1}{2}}}{H_1^{\frac{3}{4}}} = \frac{N_2 Q_2^{\frac{1}{2}}}{H_2^{\frac{3}{4}}}$$

and is called the specific speed, N_s. Although the pump specific speed is usually described as being dimensionless, it has the dimensions of $L^{3/4}T^{-3/2}$. The numerical value of the specific speed corresponds to the type of pump for required conditions of rotational speed, flowrate and head. As a guide

$N_s<0.36$	Multi-stage centrifugal/positive displacement
$0.36<N_s<1.10$	Single-stage centrifugal (radial)
$1.10<N_s<3.60$	Mixed flow pumps
$3.60<N_s<5.50$	Axial flow pumps

The maximum efficiency for each pump type spans a narrow range of specific speeds with little overlap between pump types. The specific speed is therefore valuable to identify the pump type for a particular application.

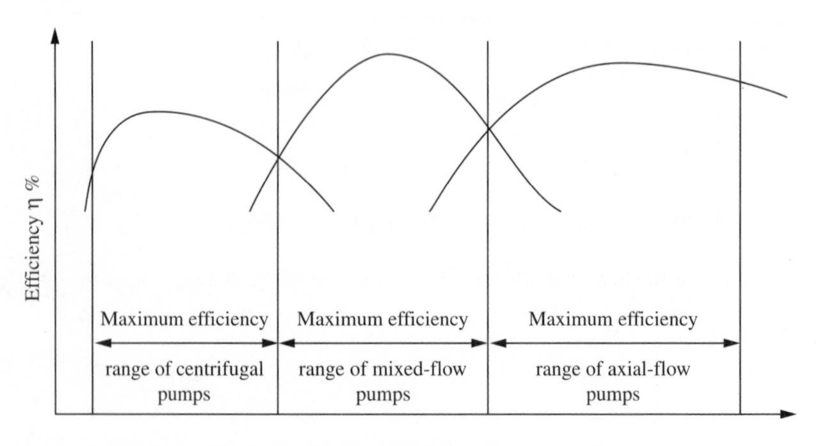

9.8 Net positive suction head: calculation 2

A low level alarm is to be installed in a vessel containing warm water at a gauge pressure of 50 kNm^{-2} to prevent cavitation at the design temperature of 40°C. Determine the minimum level of the alarm above the centre line of the pump if separate tests show that the minimum NPSH for the pump is given by

$$\text{NPSH} = 2.8 N_s^{\frac{4}{3}} H$$

where N_s is the specific speed of the pump at maximum efficiency and has a value of 0.14 (m$^{3/4}$s$^{-3/2}$), and H is the differential head and is 30 m. The frictional losses in the suction line amount to 0.2 m.

Solution

The minimum net positive suction head is

$$\begin{aligned} \text{NPSH} &= 2.8 N_s^{\frac{4}{3}} H \\ &= 2.8 \times 0.14^{\frac{4}{3}} \times 30 \\ &= 6.11 \text{ m} \end{aligned}$$

The NPSH is given by

$$\text{NPSH} = \frac{p_1 - p_v}{\rho g} + z_1 - H_L$$

where p_v is the vapour pressure of the water at 40°C (7380 Nm^{-2}) and density, ρ, is 992 kgm^{-3}. Rearranging

$$\begin{aligned} z_1 &= \text{NPSH} + \frac{p_v - p_1}{\rho g} + H_L \\ &= 6.11 + \frac{(7.38 - 50) \times 10^3}{992 \times g} + 0.2 \\ &= 1.93 \text{ m} \end{aligned}$$

The position of the low level alarm is 1.93 m above the centre line of the pump.

9.9 Effect of reduced speed on pump characteristic

The characteristic of a centrifugal pump is given by a manufacturer in dimensional form as

$$H = 40 - 140Q - 4200Q^2$$

where Q is the flowrate ($m^3 s^{-1}$) and H is the head (m). Obtain an expression for the pump characteristic for the same pump operating at 75% of its test speed.

Solution

A centrifugal pump can operate over a range of flowrates corresponding to a range of delivered head, known as the pump characteristic. The shape of the curve depends on the layout of the impeller, blades and volute. In this case, the specific speed of the pump at full operating, N, and reduced speed, N_1, is

$$N_1 = 0.75 N$$

The head coefficient for the pump in both cases is therefore

$$C_H = \frac{gH}{N^2 D^2} = \frac{gH_1}{N_1^2 D^2}$$

(see Problem 4.3, page 104)

Therefore

$$H_1 = \frac{HN_1^2 D^2}{N^2 D^2} = \frac{H(0.75N)^2 D^2}{N^2 D^2} = 0.5625 H$$

The capacity coefficient is

$$C_Q = \frac{Q}{ND^3} = \frac{Q_1}{N_1 D^3}$$

Therefore

$$Q_1 = \frac{QN_1D^3}{ND^3}$$

$$= \frac{Q(0.75N)D^3}{ND^3}$$

$$= 0.75Q$$

The pump characteristic for the test conditions is therefore

$$\frac{H_1}{0.5625} = 40 - 140 \times \frac{Q_1}{0.75} - 4200 \times \left(\frac{Q_1}{0.75}\right)^2$$

That is

$$H_1 = 22.5 - 105Q_1 - 4200Q_1^2$$

The pump characteristic for both full operating and test conditions is therefore

Discharge Q, m^3s^{-1}	0	0.01	0.02	0.03	0.04	0.05	0.06	0.08
Diff. head H, m	40.00	38.18	35.52	32.02	27.68	22.50	17.88	1.92
Diff. head H_1, m	22.50	21.03	18.72	15.57	11.58	6.75	2.13	–

Both delivered head and discharge are therefore reduced at the lower impeller speed. Centrifugal pumps usually operate with a fixed speed. However, mechanically worn pumps may operate at reduced speeds and may therefore be unable to perform their required duty.

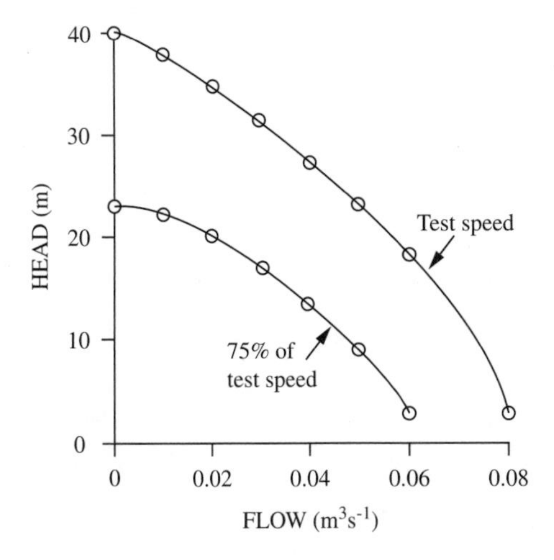

9.10 Duty point and reduced speed of a centrifugal pump

A centrifugal pump with an impeller speed of 1200 rpm has the following characteristic:

Discharge Q, m^3s^{-1}	*0.000*	*0.002*	*0.004*	*0.006*	*0.008*	*0.010*
Head H, m	*40.0*	*39.5*	*38.0*	*35.0*	*30.0*	*20.0*

The pump is used to deliver a process liquid along a pipe of length 100 m and inside diameter 50 mm and discharges from the end of the pipe at atmospheric pressure at an elevation of 10 m above the open feed tank. Determine the duty point and the speed of the pump that would result in a reduction in flow of 25%. Assume a friction factor of 0.005 for the pipe and neglect entrance and exit losses.

Solution

The duty point is the maximum flowrate that the pump can achieve for a particular power demand and forms the intersection between the pump and system characteristic curve (a plot of discharge against head). The system characteristic is usually parabolic, since the frictional head loss is proportional to the square of the fluid flowrate. Applying the Bernoulli equation, the head to be developed by the pump is

$$H = \frac{v^2}{2g} + z_2 - z_1 + H_L$$

$$= \frac{v^2}{2g} + z_2 - z_1 + \frac{4fL}{d}\frac{v^2}{2g}$$

or in terms of flowrate

$$H = z_2 - z_1 + \frac{16Q^2}{2g\pi^2 d^4}\left(\frac{4fL}{d} + 1\right)$$

$$= 10 + \frac{8 \times Q^2}{g \times \pi^2 \times 0.05^4}\left(\frac{4 \times 0.005 \times 100}{0.05} + 1\right)$$

$$= 10 + 5.42 \times 10^5 Q^2$$

The pump and system characteristic are therefore

Discharge Q, m^3s^{-1}	0.000	0.002	0.004	0.006	0.008	0.010
Head (pump), m	40.0	39.5	38.0	35.0	30.0	20.0
Head (system), m	10.0	12.2	18.7	29.5	44.7	64.2

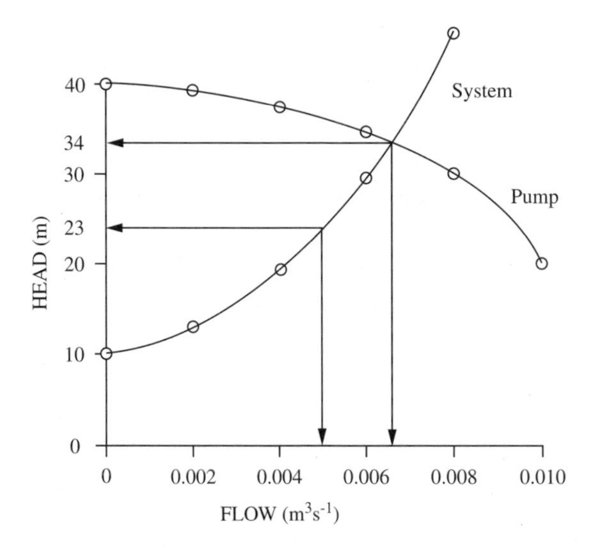

From the graph, the duty point corresponds to a flowrate of 0.0066 m^3s^{-1} and a head of 34 m. A 25% reduction in flow is therefore 0.00495 m^3s^{-1} with a head of 23 m. Since the head coefficient for the pump at the two speeds is

$$C_H = \frac{gH_1}{N_1^2 D^2} = \frac{gH_2}{N_2^2 D^2}$$

(see Problem 4.3, page 104)

then the reduced speed is

$$N_2 = N_1 \left(\frac{H_2}{H_1} \right)^{\frac{1}{2}} = 1200 \times \left(\frac{23}{34} \right)^{\frac{1}{2}} = 987 \text{ rpm}$$

The speed of the pump to reduce the flow by 25% is 987 rpm.

9.11 Power, impeller diameter, speed and delivered head

A centrifugal pump is used to transfer water from a vessel at an absolute pressure of 50 kNm^{-2} to another vessel at an absolute pressure of 150 kNm^{-2}. The elevation of the water in each vessel above the centre line of the pump is 5 m and 15 m, respectively. The connecting pipework on the suction and delivery sides of the pump has equivalent lengths of 5 m and 20 m, respectively. In addition, there is a control valve in the line which has a loss of 25% of the pump differential head; the flow controller maintains the flow at a rate of 0.05 m^{3}s^{-1}. Data from a similar geometry pump with an impeller diameter of 0.15 m and a rotational speed of 1500 rpm are

Discharge Q, ls^{-1}	*0*	*5*	*10*	*15*	*20*	*25*
Differential head H, m	*9.25*	*8.81*	*7.85*	*6.48*	*4.81*	*2.96*
Power input P, kW	*–*	*0.96*	*1.03*	*1.19*	*1.26*	*1.45*

For a liquid velocity of 2 ms^{-1}, determine the differential head, the impeller diameter and rotational speed required for operation at maximum efficiency, and the power input to the pump. Assume a friction factor of 0.005.

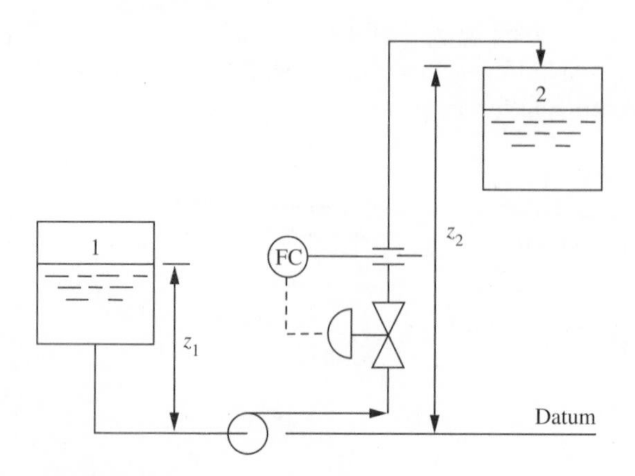

Solution

Applying the Bernoulli equation at point 1 and point 2

$$\frac{p_1}{\rho g} + z_1 + H = \frac{p_2}{\rho g} + \frac{v_2^2}{2g} + z_2 + H_L$$

where H_L is the loss across the valve and frictional loss in the pipeline for which the diameter is

$$d = \left(\frac{4Q}{\pi v}\right)^{\frac{1}{2}}$$

$$= \left(\frac{4 \times 0.05}{\pi \times 2}\right)^{\frac{1}{2}}$$

$$= 0.178 \text{ m}$$

The suction line losses are

$$H_{f(s)} = \frac{4fL_s}{d}\frac{v^2}{2g}$$

$$= \frac{4 \times 0.005 \times 5}{0.178} \times \frac{2^2}{2g}$$

$$= 0.114 \text{ m}$$

and the delivery line losses are

$$H_{f(d)} = \frac{4fL_d}{d}\frac{v^2}{2g}$$

$$= \frac{4 \times 0.005 \times 20}{0.178} \times \frac{2^2}{2g}$$

$$= 0.458 \text{ m}$$

Rearranging the Bernoulli equation

$$H = \frac{p_2 - p_1}{\rho g} + \frac{v_2^2}{2g} + z_2 - z_1 + H_{f(s)} + H_{f(d)} + 0.25H$$

The differential head is therefore

$$H = \frac{1}{1 - 0.25}\left(\frac{p_2 - p_1}{\rho g} + \frac{v_2^2}{2g} + z_2 - z_1 + H_{f(s)} + H_{f(d)}\right)$$

$$= \frac{1}{1 - 0.25} \times \left(\frac{(1.5 - 0.5) \times 10^5}{1000 \times g} + \frac{2^2}{2g} + 15 - 5 + 0.114 + 0.458\right)$$

$$= 27.96 \text{ m}$$

The actual power input for the model pump is

$$P = \frac{\rho g Q H}{\eta}$$

Rearranging, the efficiency is therefore

$$\eta = \frac{\rho g Q H}{P}$$

and calculated from the pump data as

Flowrate Q, ls^{-1}	0	5	10	15	20	25
Differential head H, m	9.25	8.81	7.85	6.48	4.81	2.96
Power input P, kW	–	0.96	1.03	1.19	1.26	1.45
Efficiency η, %	0	45	75	80	75	50

The maximum efficiency is therefore 80% for which the speed is determined from the pump head and capacity coefficients (see Problem 9.7, page 249)

$$N = N_1 \left(\frac{H}{H_1} \right)^{\frac{3}{4}} \left(\frac{Q_1}{Q} \right)^{\frac{1}{2}}$$

$$= \frac{1500}{60} \times \left(\frac{27.96}{6.48} \right)^{\frac{3}{4}} \left(\frac{0.015}{0.05} \right)^{\frac{1}{2}}$$

$$= 41 \text{ rps}$$

The diameter of the pump impeller is therefore

$$D = D_1 \left(\frac{Q N_1}{Q_1 N} \right)^{\frac{1}{3}}$$

$$= 0.15 \times \left(\frac{0.05 \times \frac{1500}{60}}{0.015 \times 41} \right)^{\frac{1}{3}}$$

$$= 0.19 \text{ m}$$

The actual power is therefore

$$P = \frac{\rho g Q H}{\eta}$$

$$= \frac{1000 \times g \times 0.05 \times 27.96}{0.80}$$

$$= 17.143 \times 10^3 \text{ W}$$

The power input is found to be 17.1 kW for a pump with an impeller of 19 cm and rotational speed of 2460 rpm. In practice, the NPSH calculation should also be completed to ensure cavitation is avoided.

Note that at maximum efficiency, the specific speed is

$$N_s = \frac{N_1 Q_1^{\frac{1}{2}}}{H_1^{\frac{3}{4}}}$$

$$= \frac{\frac{1500}{60} \times 0.015^{\frac{1}{2}}}{6.48^{\frac{3}{4}}}$$

$$= 0.75 \ (\text{m}^{3/4}\text{s}^{-3/2})$$

corresponding to a radial single-stage centrifugal pump (see Problem 9.7, page 249).

9.12 Suction specific speed

A centrifugal pump is to deliver 50 m^3h^{-1} of a process liquid. If the total head at the best efficiency point for the pump is 5 m, determine the suction specific speed as a dimensionless number for an impeller speed of 2000 rpm.

Solution

It has been the practice to define the specific speed for a pump in terms of its cavitation characteristics so that an impeller diameter D, rotating with a speed N, and delivery volumetric flowrate Q, at optimal efficiency would produce the same pump inlet conditions for a geometrically similar pump under similar operating conditions. Thus, the head coefficient C_H and capacity coefficient C_Q for the two geometrically similar pumps are

$$C_H = \frac{gH}{N^2D^2}$$

(see Problem 4.3, page 104)

$$C_Q = \frac{Q}{ND^3}$$

Eliminating D and rearranging in terms of the impeller speed

$$N = \frac{C_Q^{\frac{1}{2}}(gH)^{\frac{3}{4}}}{C_H^{\frac{3}{4}}Q^{\frac{1}{2}}}$$

Defining the suction specific speed S_n as the dimensionless group

$$S_n = \frac{C_Q^{\frac{1}{2}}}{C_H^{\frac{3}{4}}}$$

$$= \frac{NQ^{\frac{1}{2}}}{(gH)^{\frac{3}{4}}}$$

where N is the impeller speed (rps), Q is the flow through the pump and H is the net positive suction head, then

$$S_n = \frac{\frac{2000}{60} \times \left(\frac{50}{3600}\right)^{\frac{1}{2}}}{(g \times 5)^{\frac{3}{4}}}$$

$$= 0.21$$

The suction specific speed provides an indication of impeller specification as well as the likelihood of cavitation where as a guide

$S_n < 0.12$	Pumps are free from cavitation. This is a useful limit for pumps used to deliver very corrosive gas-free liquids where materials of construction which are resistant to the destructive effects of cavitation cannot be used (see Problem 9.4, page 244).
$0.12 < S_n < 0.4$	The pump inlet bore may be the same as the bore of the pump inlet connection. An inlet reducer is therefore not required. It is also recommended that bends connected to the inlet have a radius in excess of five times the inlet bore.
$0.4 < S_n < 0.7$	Materials of construction should be resistant to corrosion as pumps operate in this range with incipient cavitation. This range is typical for commercial pumps handling clean liquids.
$S_n > 0.7$	Applies only to pumps which are used to handle clean non-corrosive liquids. As cavitation is likely, it is essential that pump suction calculations and pump layout are examined closely. Inlet pipes should be straight and free from obstructions, and inlet isolation valves (such as ball valves) should not restrict flow.

9.13 Reciprocating pumps

Describe, with the aid of a diagram, the operation of reciprocating pumps, their applications and limitations.

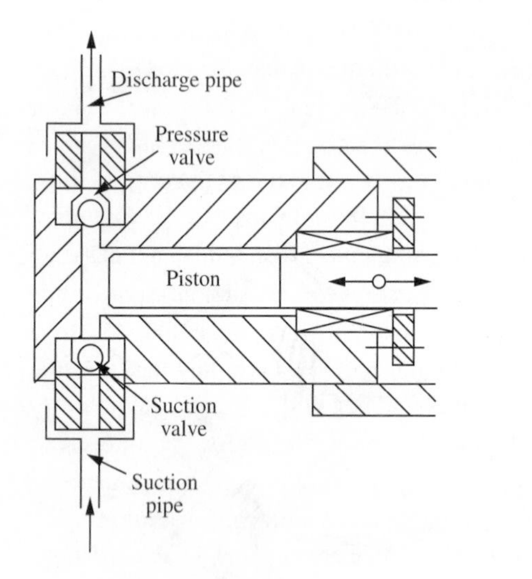

Solution

In general, reciprocating pumps transport fluid by the action of a piston on the fluid contained within a cylinder. The piston is provided with a reciprocating motion by means of a connecting rod and crank where the discharge is dependent on the swept volume and the stroke frequency; the swept volume is the amount of fluid displaced on each stroke depending on the cross-sectional area of the cylinder and the length of stroke. Check valves are needed to ensure that flow is in the correct direction.

Although they account for a relatively small proportion of all the pumps used on process plants, reciprocating pumps are applied universally across a wide range of industries. They are particularly useful for delivering fluids into high pressure process streams. Indeed, reciprocating pumps are capable of delivering heads up to several thousand bar and are limited only by the mechanical strength of the pump and slight leakage. They are also self-priming. The flow is not constant but pulsed, although the use of gas (usually nitrogen) vessels and other pulsation dampers smoothes the flow. Reciprocating pumps provide unsatisfactory performance with very viscous liquids due to slow check valve action, and with suspended solids due to attrition between the piston and cylinder.

The volumetric efficiencies of reciprocating pumps are generally high and dependent on the stroke frequency and pressure — limited largely by check valve leakage caused by wear, corrosion or mis-seating due to trapped particulates. Another limitation is the phenomenon of separation in which the pressure at the face of the piston falls below the vapour pressure of the liquid. The liquid therefore becomes separated from the piston by vapour. Further problems can arise involving vapour or gas which enters the cylinder space from around a leaking piston seal. This results in a fall in the rate of discharge, audible noise and damage to the pump.

In addition to air vessels on both the delivery and suction lines, the flow characteristic can be made smoother by the use of multi-cylinder pumps. The total flow is thus the sum of the instantaneous flows from the cylinders where odd numbers of cylinders (three or triplex, or five or quintuplex) produce the smallest pulsations (see Problem 9.15, page 265).

Abrasive, toxic and corrosive liquids, including slurries and radioactive fluids, can be handled by a special type of reciprocating pump, the diaphragm pump, which consists of a flexible diaphragm fixed at the edges and flexed to and fro. The diaphragm pump, however, requires priming prior to operation.

Another variation is the cylinder or ram pump in which a plunger replaces the piston and has a clearance between the cylinder wall. It is more expensive than the piston pump although leakage is easier to detect, the packing is easier to replace and it is more suitable for delivering high heads.

9.14 Single-acting reciprocating piston

A single-acting reciprocating piston has a piston area of 100 cm² and a stroke of 30 cm and is used to lift water a total height of 15 m. If the speed of the piston is 60 rpm and the actual quantity of water lifted is 10.6 m^3h^{-1}, determine the coefficient of discharge and the theoretical power required to drive the pump.

Solution

The expected volume displaced in the piston cylinder is

$$V = ALN$$

$$= 0.01 \times 0.3 \times \frac{60}{60}$$

$$= 3 \times 10^{-3}\ \text{m}^3\text{s}^{-1}$$

The actual volume discharged is

$$V_{act} = \frac{10.6}{3600}$$

$$= 2.94 \times 10^{-3}\ \text{m}^3\text{s}^{-1}$$

The coefficient of discharge is therefore

$$C_d = \frac{V_{act}}{V}$$

$$= \frac{2.94 \times 10^{-3}}{3 \times 10^{3}}$$

$$= 0.98$$

and the power required is

$$P = \rho g Q H$$

$$= 1000 \times g \times \frac{10.6}{3600} \times 15$$

$$= 433\ \text{W}$$

The coefficient of discharge and theoretical power required are 0.98 and 433 W, respectively.

9.15 Discharge from reciprocating pumps

Determine the ratio of peak to average discharge from single, two, three, four and five-cylinder reciprocating pumps.

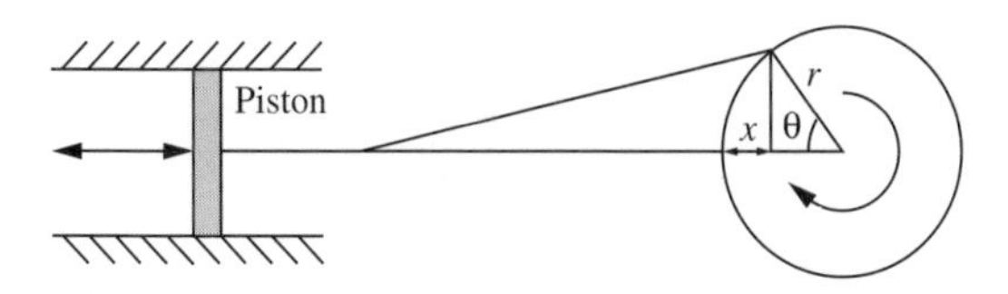

Solution

It is assumed that the piston in each of these pumps operates with perfect simple harmonic motion and that the volumetric efficiency and acceleration effects of the fluid are ignored. Thus, for a single-cylinder pump, the displacement of liquid is

$$x = r - r\cos\theta$$

$$= r(1 - \cos\theta)$$

$$= r(1 - \cos\omega t)$$

The velocity of the piston is

$$v = \frac{\mathrm{d}x}{\mathrm{d}t}$$

$$= \omega r \sin\omega t$$

$$= \omega r \sin\theta$$

The volume displaced on the pressure stroke (between 0 and π) is therefore

$$V = A\omega r\int_0^{\pi} \sin\theta d\theta$$

$$= -A\omega r\cos(\pi - 0)$$

$$= 2A\omega r$$

where A is the area of the piston. The average flow over the cycle (between 0 and 2π) is therefore

$$Q = \frac{V}{2\pi}$$

$$= \frac{2A\omega r}{2\pi}$$

$$= \frac{A\omega r}{\pi}$$

Since the peak flow occurs at $\pi/2$, then

$$Q_{peak} = A\omega r \sin\frac{\pi}{2}$$

$$= A\omega r$$

The ratio of peak to average flow is therefore

$$\frac{Q_{peak}}{Q} = \frac{A\omega r}{\frac{A\omega r}{\pi}}$$

$$= \pi$$

$$= 3.14$$

Likewise, for duplex (two-cylinder), triplex (three-cylinder), quadruplex (four-cylinder) and quintruplex (five-cylinder) pumps, the ratios are found to be 1.57, 1.05, 1.11 and 1.02, respectively.

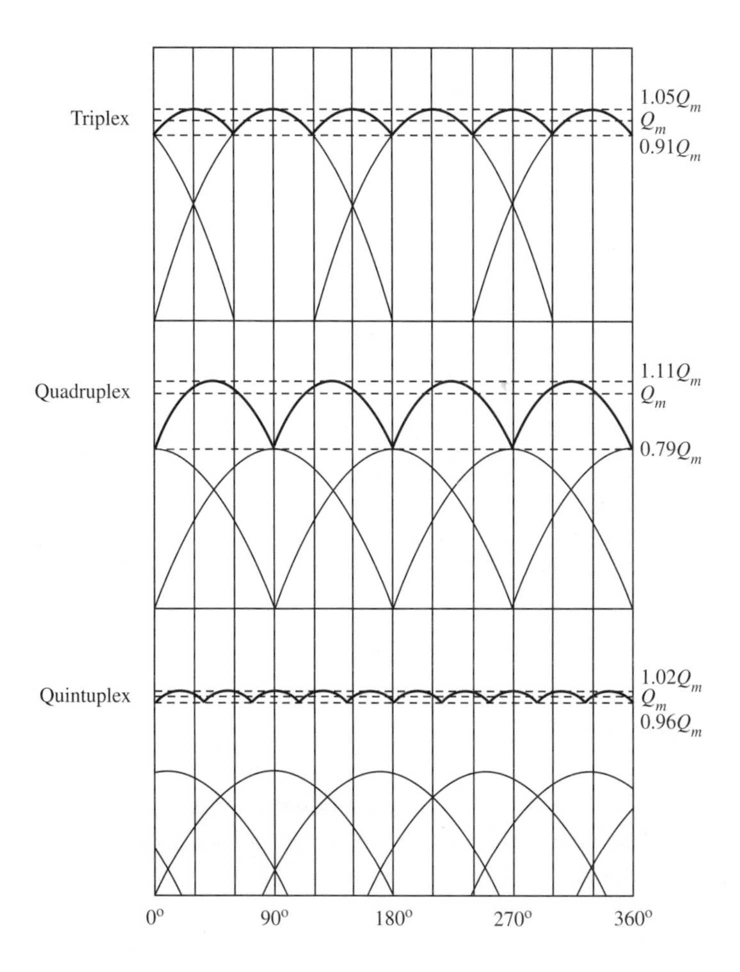
Triplex
1.05Q_m
Q_m
0.91Q_m
Quadruplex
1.11Q_m
Q_m
0.79Q_m
Quintuplex
1.02Q_m
Q_m
0.96Q_m
0°
90°
180°
270°
360°

9.16 Rotary pumps

Describe briefly, with a labelled diagram, the layout and operation of rotary pumps.

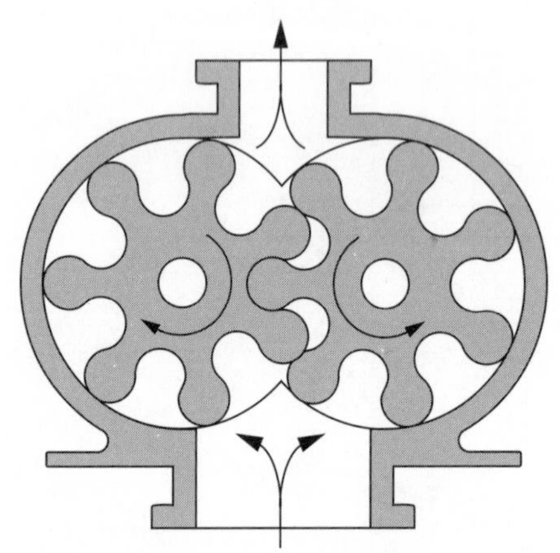

Rotary pump (gear type)

Solution

In rotary pumps, portions of liquid are carried from the entry port to the delivery port in compartments within the pump formed between the rotor and the stator. They are manufactured to a very close tolerance to reduce back-leakage. They are therefore expensive to fabricate. The flowrate depends on the size and number of compartments as well as the frequency of rotation. They have the capability of delivering against high heads, are self-priming and provide smooth and continuous flow. They offer higher speeds than reciprocating pumps and are suitable for viscous liquids since no check valves are required. But they require careful maintenance and are unsuitable for liquids without lubricating properties such as water, and for slurries due to small clearances and possibility of friction.

There are various forms of rotary pump. The Mono pump, for example, has a helical worm rotating inside a similarly shaped flexible stator. This type of pump can handle corrosive and gritty liquids, and is commonly used to feed slurries to filter presses, but it must never be allowed to run dry. The Roots blower is used for transporting large volumes of gas at low gauge pressure, using rotors in the form of cycloidal lobes.

Note that it is essential to provide all positive displacement pumps (reciprocating and rotary types) with a safety relief valve to avoid excessive pressure build-up in the event of the delivery line being inadvertently closed or blocked.

Further problems

(1) Describe briefly, with the use of clearly labelled sketches, the essential features, uses and limitations of rotary pumps.

(2) Explain the phenomenon of cavitation applied to centrifugal pumps and describe practical measures which can lead to its reduction.

(3) Construct a flowchart for the decision-making process for the selection of centrifugal pumps.

(4) Determine the additional available NPSH of a system for a pump used to transfer water that allows for worst case meteorological conditions.

Answer: 0.62 m

(5) Describe the essential features of the centrifugal pump; include a sketch.

(6) A centrifugal pump is used to deliver a dilute solution of sodium hydroxide with a density of 1050 kgm^{-3} from a large open buffer tank to a header tank 10 m above it. The velocity of the solution in the suction pipe to the pump is 1.5 ms^{-1}. The suction pipe has an internal diameter of 3 cm and the delivery pipe has an internal diameter of 4 cm. If the frictional head loss is 1.5 m, determine the theoretical power requirement for the pump.

Answer: 126 W

(7) Explain the significance of net positive suction head (NPSH).

(8) Sulphuric acid with a specific gravity of 1.83 and a viscosity 0.04 Nsm^{-2} is pumped into a large vessel through a 25 cm inside diameter pipe 30 m long inclined at an angle of 30° to the horizontal. Determine the power required to pump 0.064 m^3s^{-1} into the vessel. A friction factor of 0.008 may be assumed.

Answer: 17.6 kW

(9) Sulphuric acid with a specific gravity of 1.84 is to be pumped from an open tank to a process column at a rate of 25 kgs^{-1}. The column operates at a pressure of 125 kNm^{-2} and the acid is sprayed from a nozzle located 20 m above the acid level in the tank at a linear velocity of 3 ms^{-1}. If the total additional losses are equivalent to 2.75 m of water head, determine the power required to pump the acid if the pump has an efficiency of 65%.

Answer: 9.01 kW

(10) 75% sulphuric acid, of density 1650 kgm^{-3} and viscosity 8.6×10^{-3} Nsm^{-2}, is pumped at a rate of 10000 kgh^{-1} through an 800 m section of smooth pipe with an internal diameter of 50 mm with a rise of 12 m. If the pump efficiency is 50%, determine the power requirement. Assume the Blasius equation applies.

Answer: 1.74 kW

(11) Liquid, of density 850 kgm^{-3}, is pumped at a rate of 10 kgs^{-1} from an open storage tank at atmospheric pressure to a distillation column which operates at a pressure of 500 kNm^{-2} above atmospheric pressure. The inside diameter of the pipe supplying the column is 10 cm, and the surface of the liquid in the tank is 5 m above the ground level and can be considered constant. The column feed point is 24 m above the ground. Determine the power required for the pump if the overall efficiency of pump and motor is 70%. Losses due to friction amount to 3 m.

Answer: 11.5 kW

(12) The overhead condensate from a distillation column of density 600 kgm^{-3} and viscosity 0.6×10^{-4} Nsm^{-2} is fed at a rate of 5 kgs^{-1} from the condensate drum held at a gauge pressure of 12 bar to a pressurized storage vessel at a gauge pressure of 10.2 bar prior to further processing. The pipeline has a length of 2.5 km and an inside diameter of 7.5 cm with a wall roughness of 0.046 mm. If the difference in level of the condensate in the condensate drum and the discharge of the pipe into the storage vessel is 1 m, determine the power requirement of the pump and motor assembly if their combined efficiency is 65%. Allow for entrance and exit losses and where the pipe contains fifty-six 90° elbows of 0.7 velocity heads each and four open gate valves of 0.15 velocity heads each.

Answer: 7.3 kW

(13) Ethane is removed from an LPG mixture in a distillation column in which the ethane-rich overhead vapour is condensed, and some is returned to the top of the column as reflux. The top of the column operates at a gauge pressure of 12 bar. The gauge pressure in the reflux drum is 11.5 bar. The difference in elevation between the top tray of the column and the level of liquid in the reflux drum is 10 m. The connecting pipework consists of 50 m of 0.254 m inside diameter pipe with 27 bends, two gate valves and one control valve. If the flowrate of refluxed liquid, of density 600 kgm^{-3}, is 299 m^3h^{-1}, determine the

power requirement of the reflux pump and motor assembly assuming an efficiency of 75%. Allow for entrance and exit losses where each bend and valve has a loss equivalent to 0.7 and 0.15 velocity heads, respectively. The constant friction factor may be taken as 0.006.

Answer: 14.4 kW

(14) An LPG mixture, of density 600 kgm^{-3} and viscosity 6×10^{-4} Nsm^{-2}, is pumped at a rate of 10 kgs^{-1} from a storage vessel held at a gauge pressure of 500 kNm^{-2} to a distillation column for separation. The pipeline is 50 m in length and has a bore of 10 cm and surface roughness of 0.046 mm. If the gauge pressure at the feed tray in the column is 1.1 MNm^{-2} and is located 20 m above the centre line of the pump while the surface of the LPG mixture in the storage vessel is 3 m above the pump, determine the power requirement of the pump and motor assembly assuming an efficiency of 60%. Allow for entrance and exit losses; the pipe has eight 90° elbows and three open gate valves of 0.7 and 0.15 velocity heads each, respectively.

Answer: 20.2 kW

(15) An ethanol distillation column operating at a gauge pressure of 50 kNm^{-2} receives a liquid feed of density 950 kgm^{-3} and viscosity 9×10^{-4} Nsm^{-2} at a rate of 39 m^3h^{-1} from an open storage vessel. The connecting pipework consists of 50 m of 10 cm inside diameter pipe with an absolute roughness of 0.04 mm. The pipe has a total of 12 elbows and three gate valves. The level of liquid feed in the storage tank is steady at 4 m above the centre line of the pump and the feed point on the distillation column is at an elevation of 5 m above the pump. If the overall efficiency of the pump and motor is 70%, determine the power requirement of the pump. Allow for entrance and exit losses and where the loss for each bend and valve is equivalent to 0.7 and 0.12 velocity heads, respectively.

Answer: 1.2 kW

(16) A methanol-water liquid mixture is to be transported from a storage tank to a distillation column for separation. The mixture, of density 895 kgm^{-3}, is stored within an enclosed tank maintained at an absolute pressure of 1 atmosphere and is to be pumped to the column at a rate of 11 kgs^{-1} through a pipe of inside diameter 12 cm. The column, which operates at a gauge pressure of 50 kNm^{-2}, has a feed point 15 m above the level of the liquid mixture in the storage tank. If the overall efficiency of the pump and motor is 65% and the losses due

to friction amount to 4 m, determine the power requirement of the pump.

Answer: 4.1 kW

(17) Show that the work exerted on a fluid by a centrifugal pump is given by

$$W = \frac{NF}{53.3}$$

where W is the work (Watts), N is the speed of the impeller (revolutions per minute) and F is the force (Newtons) on a torque arm measured from the centre line of the shaft which has a radius of 179 mm.

(18) Experimental data were obtained from laboratory tests of a centrifugal pump operating at 2500 rpm using water as the test fluid. F is the force exerted on a torque arm measured from the centre line of the shaft which has a radius of 179 mm.

Flowrate, Q ($\times 10^{-3}$ m^{3}s^{-1})	1.47	1.31	1.18	1.00	0.91	0.71	0.48
Diff. pressure, Δp (kNm^{-2})	18.8	36.0	44.8	61.5	63.6	71.5	79.7
Force on arm, F (N)	4.6	4.1	4.0	3.7	3.3	3.0	2.4

Determine the flowrate delivered at maximum pump efficiency. Comment on the general performance of the pump.

(19) Water is to be transported from one tank to another using a 10 m pipeline of inside diameter 25 mm. A centrifugal pump is to be used to transport the water and is positioned below the first tank whose free surface is 2 m above the freely discharging outlet. Determine the duty point for the pump characteristic given below. The friction factor may be assumed to be 0.005. Allow for entrance and exit losses.

Discharge, m^{3}min^{-1}	0.000	0.008	0.012	0.018	0.024
Head, m	3.5	3.2	3.0	2.4	1.4

Answer: 0.02 m^{3}min^{-1}

(20) A centrifugal pump is used to transfer a liquid of density 790 kgm^{-3} from an open storage vessel. The centre of the pump is located at an elevation of 2 m above the liquid in the vessel. Determine the available NPSH if the vapour pressure of the liquid at the temperature of operation is 25 kNm^{-2} and the frictional

head in the suction pipe is 0.3 m. Allow for worst case meteorological conditions.

Answer: 6.77 m

(21) A liquid hydrocarbon liquid mixture, of density 538 kgm^{-3} and viscosity 3.04×10^{-4} Nsm^{-2}, is pumped from the bottom of a distillation column operating at 120 kNm^{-2} through a reboiler at a rate of 14×10^{3} kgh^{-1}. The centre line of the pump is a vertical distance H below the column take-off point and the suction line has a total length of $H+3$ metres and an inside diameter of 65 mm. The reboiler is at an elevation of $H+0.5$ metres above the pump. If the pressure drop through the reboiler is 14 kNm^{-2}, determine the differential head across the pump. Neglect kinetic energy head loss and losses due to fittings, and entrance and exit losses. Assume a constant Fanning friction factor of 0.004. The NPSH for the pump is 3 m.

Answer: 30.11 m

(22) A centrifugal pump delivers water at a rate of 0.02 m^3s^{-1} against a head of 10 m. If the impeller speed is 1500 rpm, determine the specific speed N_s.

Answer: 0.63 $m^{3/4}s^{-3/2}$

(23) Determine the suction specific speed S_n for a centrifugal pump operating at an impeller speed of 40 rps delivering a flow of 110 m^3h^{-1} against a head of 14 m.

Answer: 0.174

(24) Determine the numerical value of k which relates the specific speed of a centrifugal pump, N_s, to suction specific speed, S_n, as

$$S_n = \frac{N_s}{k}$$

for a pump with a discharge measured in litres per minute.

Answer: 176

30

Glossary of terms

Absolute pressure
Pressure measured above a vacuum.

Absolute roughness
Average height of undulations and imperfections on the inner surface of a pipe wall.

Angular velocity
Rotational speed expressed in radians per second.

Archimedes' principle
States that the upthrust on a partially or totally immersed body is equal to the weight of liquid displaced.

Average velocity
Defined as the total flowrate per unit flow area.

Boundary layer
Region between a wall and a point in the flowing fluid where the velocity is at a maximum.

Cavitation
Destructive collapse of vapour in liquid in regions of high pressure.

Centrifugal pump
Mechanical device used to transport fluids by way of an enclosed rotating impeller imparting velocity to the fluid.

Coefficient of contraction
Ratio of area of *vena contracta* to orifice area.

Coefficient of discharge
Ratio of actual flowrate to theoretical flowrate through an opening or restriction.

Coefficient of velocity
Ratio of actual to theoretical velocity for a discharging jet of fluid.

Density
Measure of the quantity of a substance per unit volume.

Dynamic viscosity
Phenomenon in which liquids shear between adjacent layers.

Dynamical similarity
Similarity between systems when applied forces have the same ratio.

Elbow
A curved 90° section of pipework.

Entrance and exit losses
Irreversible energy loss when fluids enter or leave pipes.

Entrance transition length
Distance between pipe entrance and point of fully developed laminar flow.

Equivalent hydraulic diameter
Used in place of diameter d in turbulent flow equations when the pipe or duct cross-section is not circular. Equal to four times the flow area divided by the wetted perimeter. Does not apply to laminar flow.

Equivalent length
The length of straight pipe that would give the same pressure drop as the fitting of the same nominal diameter expressible in pipe diameters of the straight pipe.

Friction factor
Empirical factor and is a function of Reynolds number and relative pipe roughness.

Fluid
A substance which offers no resistance to change of shape by an applied force.

Forced vortex
Circular stream of liquid the whirl of which is caused by an external source such as a stirrer or impeller.

Free jet
Discharging liquid from a pipe or orifice which is no longer influenced by the applied head.

Free surface
Interface of a liquid with a vapour or gas above.

Free vortex
Streamlines that move freely in horizontal concentric circles with no variation of the total energy across them.

Frictional head
Head required by a system to overcome resistance to flow in pipes and associated fittings.

Gate valve
Device used to regulate flow in a pipe, consisting of a vertical moving section across the flow area.

Gauge pressure
Pressure measured above atmospheric.

Geometrical similarity
Systems which have similar physical dimensions such that their ratio is a constant.

Head
Pressure of a liquid expressed as the equivalent height a column of the liquid would exert.

Laminar flow
Flow that occurs when adjacent layers of fluid move relative to one another in smooth streamlines. Flow occurs in pipes for Reynolds numbers below 2000.

Mass flowrate
Mass flow of a fluid per unit time.

Mass loading
Rate of mass flow per unit area.

Mean hydraulic depth
Ratio of flow area to wetted perimeter in an open channel.

Net positive suction head (NPSH)
Total suction head determined at the suction nozzle of a centrifugal pump less the vapour pressure.

NPSHA
NPSH available is the absolute pressure at the inlet of a centrifugal pump and is a function of elevation, temperature and pressure.

NPSHR
The NPSH required by a centrifugal pump is a function of the pump characteristics and must be exceeded by NPSHA.

Newtonian fluids
Class of fluids in which viscosity is independent of shear stress and time.

Non-Newtonian liquids
Liquids in which the viscosity depends on shear stress and/or time.

No-slip condition
Assumption that the fluid layer in contact with a surface is stationary.

Notch
Rectangular or triangular cut in a weir.

Open channel
Conduit carrying liquid with a free surface.

Orifice
Small hole or opening.

Orifice meter
Device used to measure flowrate by the pressure drop through a small hole.

Piezometer
Device used to measure pressure, consisting of an open vertical (glass) tube attached to the side of a horizontal pipe.

Pipeline
Long section of large bore pipe.

Pitot tube
Device used to measure velocity head of a flowing fluid.

Power
Work done per unit time.

Pressure
Force of a fluid applied over a given area.

Pressure differential
Difference in pressure between two points.

Pressure head
Pressure-volume energy of a fluid expressed in head form.

Relative roughness
Ratio of absolute pipe wall roughness to pipe inside diameter in consistent units.

Reynolds number
Dimensionless number expressing the ratio of inertial to viscous forces in a flowing fluid.

Rotameter
Instrument used to measure flowrate by the vertical elevation of a float in a tapered tube.

Separation
Phenomenon of fluid streamlines changing direction due to changes in boundary shape as a result of fluid inertia or velocity distribution near the boundary surface.

Separator
Vessel used to separate two liquids, usually as layers.

Shear stress
Force of a liquid applied over a surface.

Shear rate
Change in velocity of a fluid perpendicular to flow.

Sluice
Submerged opening.

Specific gravity
Ratio of mass of a liquid to mass of water occupying the same volume.

Specific speed
Dimensionless group used to classify centrifugal pump impellers at optimal efficiency with respect to geometric similarity.

Static head
Potential energy of a liquid expressed in head form.

Steady flow
No change of fluid velocity with respect to time.

Streamline
An imaginery line which lies in the direction of flow.

Siphon
Transfer of liquid from one vessel to another at a lower elevation by means of a pipe whose highest point is above the surface of the liquid in the upper vessel.

Transition flow
Flow regime between laminar and turbulent flow ($2000<Re<4000$). Velocity fluctuations may be present and impossible to predict.

Turbulent flow
Flow regime characterized by fluctuating motion and erratic paths above Reynolds numbers of 4000. Flow occurs when inertial forces predominate resulting in macroscopic mixing of the fluid.

Uniform flow
No change in fluid velocity at a given time with respect to distance.

Velocity gradient
See 'Shear rate'.

Velocity head
Kinetic energy of a fluid expressed in head form.

Vena contracta
Minimum flow area formed downstream by a fluid flowing through a sharp-edged opening.

Venturi meter
Device used to measure flowrate by the pressure drop through a tapered section of tube or pipe.

Viscosity
See 'Dynamic viscosity'.

Viscous flow
See 'Laminar flow'.

Volute chamber
Spiral casing of centrifugal pump surrounding impeller.

Weir
Vertical obstruction across a channel.

Weir crest
Top of a weir over which a liquid flows.

Selected recommended texts

Acheson, D.J., 1996, *Elementary Fluid Dynamics*, 4th edition (Oxford University Press, Oxford).

Boxer, G., 1988, *Work out Fluid Mechanics* (MacMillan Press Ltd, London).

Currie, I.G., 1993, *Fundamentals of Fluid Mechanics*, 2nd edition (McGraw-Hill, New Jersey).

Darby, R., 1996, *Chemical Engineering Fluid Dynamics* (Marcel Dekker).

Douglas, J.F., Gasiorek, J.M. and Swaffield, J.A., 1996, *Fluid Mechanics*, 3rd edition (Longman, Harlow).

Douglas, J.F. and Mathews, R.D., 1996, *Solving Problems in Fluid Mechanics*, Volumes 1 and 2, 3rd edition (Longman, Harlow).

Fox, R.W. and McDonald, A.T., 1992, *Introduction to Fluid Mechanics*, 4th edition (John Wiley, New York).

French, R.H., 1994, *Open Channel Hydraulics* (McGraw-Hill, New Jersey).

Granet, I., 1996, *Fluid Mechanics*, 4th edition (Prentice Hall, New Jersey).

Holland, F.A. and Bragg, R., 1995, *Fluid Flow for Chemical Engineers* (Edward Arnold, London).

Hughes, W.F. and Brighton, J.A., 1991, *Schuam's Outline of Theory and Problems of Fluid Dynamics*, 2nd edition (McGraw-Hill, New Jersey).

Ligget, J.A., 1994, *Fluid Mechanics* (McGraw-Hill, New York).

Massey, B.S., 1989, *Mechanics of Fluids*, 6th edition (Chapman and Hall, London).

Mott, R.L., 1994, *Applied Fluid Mechanics*, 4th edition (Prentice Hall, New Jersey).

Munson, B.R., Young, D.F. and Okiishi, T.H., 1994, *Fundamentals of Fluid Mechanics*, 2nd edition (John Wiley, New York).

Pnueli, D. and Gutfinger, C., 1997, *Fluid Mechanics* (Cambridge University Press, Cambridge).

Roberson, J.A. and Crowe, C.T., 1997, *Engineering Fluid Mechanics*, 6th edition (John Wiley, New York).

Sharpe, G.T., 1994, *Solving Problems in Fluid Dynamics* (Longman, Harlow).

Street, R.L., Watters, G.Z. and Vennard, J.K., 1996, *Elementary Fluid Mechanics*, 7th edition (John Wiley, New York).

Streeter, V.L. and Wylie, E.B., 1985, *Fluid Mechanics*, 8th edition (McGraw-Hill Book Co, London).

Tritton, D.J., 1997, *Physical Fluid Dynamics*, 2nd edition (Oxford Series Publication, Oxford).

Turner, I.C., 1996, *Engineering Applications of Pneumatics and Hydraulics* (Edward Arnold, London).

Vardy, A., 1990, *Fluid Principles* (McGraw-Hill Book Co, London).

Widden, M., 1996, *Fluid Mechanics* (Macmillan Press Ltd, London).

White, F.M., 1994, *Fluid Mechanics*, 3rd edition (McGraw-Hill, New York).

Young, D.F., Munson, B.R. and Okiishi, T.H., 1997, *A Brief Introduction to Fluid Mechanics* (John Wiley, New York).

Nomenclature and preferred units

Quantity	Name	Symbol	Type of SI unit	In terms of base SI
Area		m^2	Base	
Degree	plane angle		Allowable	$\pi/180$ rad
Density		kgm^{-3}	Base	
Force	Newton	N	Derived	$kgms^{-2}$
Length	metre	m	Base	
Length	kilometre	km	Allowable	10^3 m
Mass	kilogram	kg	Base	
Mass flow		kgs^{-1}	Base	
Mass loading		$kgm^{-2}s^{-1}$	Base	
Pressure	bar	bar	Allowable	$10^5\ kgm^{-1}s^{-2}$
Pressure	Pascal	Pa	Derived	$kgm^{-1}s^{-2}$
Pressure		Nm^{-2}	Derived	$kgm^{-1}s^{-2}$
Radian	plane angle	rad	Supplementary	
Revolution	angular displacement	rps	Allowable	2π rad
Shear stress		Nm^{-2}	Derived	$kgm^{-1}s^{-2}$
Surface tension		Nm^{-1}	Derived	kgs^{-2}
Time	hour	h	Allowable	3600 s
Time	minute	min	Allowable	60 s
Time	second	s	Base	
Torque		Nm	Derived	kgm^2s^{-2}
Velocity		ms^{-1}	Base	
Viscosity		Nsm^{-2}	Derived	$kgm^{-1}s^{-1}$
Volume		m^3	Base	
Volume	litre	l	Allowable	$10^{-3}\ m^3$
Volume flow		m^3s^{-1}	Base	
Work	joule	J	Derived	kgm^2s^{-2}
Watt	power	W	Derived	kgm^2s^{-3}

Useful conversion factors

Quantity	Multiplier	SI or preferred unit
Acceleration		
cms^{-2}	0.01	ms^{-2}
fts^{-2}	0.3048	ms^{-2}
Area		
mm^2	0.000001	m^2
cm^2	0.0001	m^2
in^2	0.0006451	m^2
ft^2	0.0929	m^2
yd^2	0.836	m^2
Density		
SG	1000	kgm^{-3}
gcm^{-3}	1000	kgm^{-3}
$lbm.ft^{-3}$	16.018	kgm^{-3}
$lbm.UKgal^{-1}$	99.77	kgm^{-3}
$lbm.USgal^{-1}$	119.83	kgm^{-3}
Energy		
kJ	1000	J
Nm	1	J
lbf-ft	1.356	J
Force		
kN	1000	N
dyne	0.00001	N
lbf	4.448	N
UKtonf	9964	N
UStonf	8896	N

Length

mm	0.001	m
cm	0.01	m
km	1000	m
in	0.0254	m
ft	0.3048	m
yd	0.9144	m

Mass

g	0.001	kg
oz(troy)	3.1103	kg
oz(avdp)	2.8349	kg
lbm	0.4536	kg

Plane angle

deg(°)	0.01745	rad

Power

kW	1000	W
Js^{-1}	1	W
$ft.lbf.s^{-1}$	1.356	W
hp	745.7	W

Pressure

Pa	1	Nm^{-2}
$kgm^{-1}s^{-2}$	1	Nm^{-2}
mmHg	133.3	
bar	100,000	Nm^{-2}
Std atm	101,300	Nm^{-2}
$dyne.cm^{-2}$	0.1	Nm^{-2}
$lbf.in^{-2}$ (psi)	6895	Nm^{-2}

Pressure drop

$Pa.m^{-1}$	1	$Nm^{-2}m^{-1}$
$psi.ft^{-1}$	0.022621	$Nm^{-2}m^{-1}$

Surface tension

kgs^{-1}	1	Nm^{-1}
$lbf.ft^{-1}$	14.59	Nm^{-1}
$dyne.cm^{-1}$	0.001	Nm^{-1}
$dyne.m^{-1}$	0.00001	Nm^{-1}

Time

min	60	s
h	3600	s
d	86,400	s

Velocity (linear)

cms^{-1}	0.01	ms^{-1}
fts^{-1}	0.3048	ms^{-1}

Velocity (angular)

rps	0.10472	$rads^{-1}$
rpm	6.2832	$rads^{-1}$
$degs^{-1}$	0.017453	$rads^{-1}$

Viscosity (dynamic)

Pa.s	1	Nsm^{-2}
$kgm^{-1}s^{-1}$	1	Nsm^{-2}
$lbf.s.in^{-2}$	6894.7	Nsm^{-2}
cP	0.001	Nsm^{-2}
P	0.1	Nsm^{-2}

Viscosity (kinematic)

ft^2s^{-1}	0.092903	m^2s^{-1}
cSt	0.000001	m^2s^{-1}
St	0.0001	m^2s^{-1}

Volume

ml	0.000001	m^3
dm^3	0.001	m^3
litre	0.001	m^3
in^3	0.0000164	m^3
ft^3	0.02831	m^3
yd^3	0.7645	m^3
bbl (42USgal)	0.1589	m^3
UKgal	0.004546	m^3
USgal	0.003785	m^3

Volume flow

m^3h^{-1}	0.0002777	m^3s^{-1}
ft^3s^{-1}	0.028317	m^3s^{-1}
$bbl.h^{-1}$	0.0004416	m^3s^{-1}
$UKgal.h^{-1}$	0.0045461	m^3s^{-1}
$USgal.h^{-1}$	0.003754	m^3s^{-1}

Physical properties of water (at atmospheric pressure)

Temperature, °C	Density, kgm^{-3}	Viscosity, Nsm^{-2}	Vapour pressure, Nm^{-2}	Surface tension, Nm^{-1}
0	1000	0.00790	611	0.076
5	1000	0.00152	872	0.075
10	1000	0.00131	1230	0.075
15	999	0.00114	1700	0.074
20	998	0.00100	2340	0.074
25	997	0.00089	3170	0.073
30	996	0.00080	4240	0.072
35	994	0.00072	8600	0.071
40	992	0.00065	7380	0.070
45	990	0.00060	9550	0.069
50	988	0.00055	12,300	0.068
55	985	0.00051	15,740	0.067
60	983	0.00047	20,000	0.067
65	980	0.00044	25,010	0.066
70	978	0.00040	31,200	0.065
75	975	0.00038	38,550	0.064
80	972	0.00035	47,400	0.063
85	969	0.00033	57,800	0.062
90	965	0.00031	70,100	0.061
95	962	0.00030	84,530	0.060
100	958	0.00028	101,300	0.059

Losses for turbulent flow through fittings and valves

Fitting or valve		Loss coefficient, k
Elbow 45° standard		0.35
Elbow 45° long radius		0.2
Elbow 90° standard		0.75
Elbow 90° long radius		0.45
Elbow 90° square		1.3
Bend 180° close return		1.8
Tee, standard, along run		0.4
Tee, standard, through branch		1.5
Pipe entry	sharp	0.5
	well rounded	0.05
	Borda	1.0
Pipe exit	sharp	0.5
Gate valve	fully open	0.15
	¾ open	0.9
	½ open	4.5
	¼ open	20.0
Diaphragm valve	fully open	2.3
	¾ open	2.6
	½ open	4.3
	¼ open	21.0
Globe valve	fully open	6.0
	½ open	9.0
Angle valve	fully open	3.5
Butterfly valve	fully open	0.5

Check valve, swing			2–3.5
Pump foot valve			1.5
Spherical plug valve	fully open		0.1
Flow meters	orifice	β=0.2	2.6
		β=0.5	1.8
		β=0.8	0.6
	nozzle	β=0.2	1.0
		β=0.5	0.55
		β=0.8	0.15
	venturi		0.15
	turbine		6.0
Cyclone			10–20

Equivalent sand roughness of pipes

Type	Equivalent sand roughness, ε (mm)
Glass	0.0003
Drawn tubing	0.0015
Steel, wrought iron	0.0457
Asphalted cast iron	0.1219
Galvanized iron, new	0.1524
Galvanized iron, old	0.2743
Cast iron	0.2591
Wood stave	0.1829
Concrete	0.3048–3.048
Riveted steel	0.9144–9.144

Manning coefficient for various open channel surfaces

Type		Manning coefficient, n ($m^{-1/3}s$)
Welded steel		0.012
Planed wood		0.012
Unplaned wood		0.013
Concrete culvert		0.013
Unfinished concrete		0.014
Cast iron		0.015
Straight sewer		0.015
Riveted steel		0.016
Galvanized iron		0.016
Earth channels	clean	0.018
	stony	0.035
	unmaintained	0.080
Natural streams	clean	0.030
	stony	0.045
	deep pools	0.100
Rivers	no boulders	0.025
	irregular	0.035

Moody plot

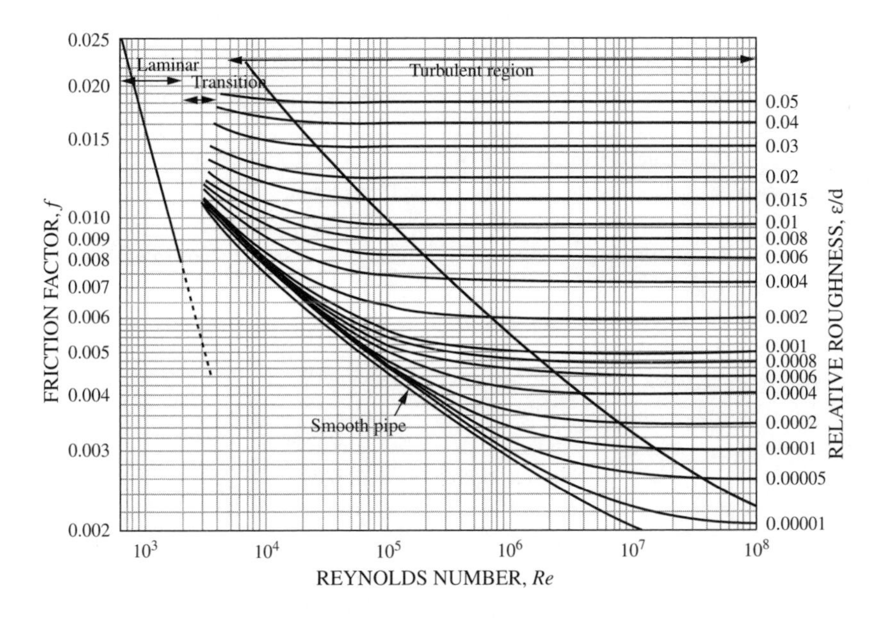

Plot of friction factor versus Reynolds number developed by Louis F. Moody.

Laminar flow ($Re<2000$)

Transition region ($2000<Re<3000$)

Turbulent region ($Re>3000$)

For rough-walled pipes

$$Re = \frac{100}{\sqrt{f}\left(\frac{\varepsilon}{d}\right)}$$

(see Problem 8.11, page 217)

Index

P